THE FORCE OF ENERGY:

A Business History of the Detroit Edison Company

Walker Lee Cisler.

THE FORCE OF ENERGY:

A Business History of the Detroit Edison Company

by

RAYMOND C. MILLER

MICHIGAN STATE UNIVERSITY PRESS
1971

Library of Congress Catalog Card Number: 70–154288
STANDARD BOOK NUMBER: 87013–164–8
MANUFACTURED IN THE UNITED STATES OF AMERICA

★
★
★
★
★

To Florence and Jane

Contents

Illustrations

The Meaning of Responsibility

AN INTRODUCTION

In almost melodramatic fashion, the issue of ultimate responsibility was raised recently in the annual stockholders meetings of two of our nation's greatest corporations. In one company, the world's largest manufacturer of automobiles, the proposal would have directed management to subordinate the property rights of the shareholders (and the payment of dividends) to the public interest in the control of air pollution. In the other company, the world's largest privately owned public utility, the proposal would prohibit all expenditures which did not contribute directly to the profit interest of the shareholders.

These extreme positions illustrate how far from agreement Americans are in their conception of corporate responsibility, and in their identification of business obligations. Even the vocabulary used in the discussion reveals the lack of common understanding. This would indicate that the corporation and its role invite an historical examination, and it is to this problem that we now address ourselves.

This volume presents the record of one of America's great business organizations, but it is not a vanity publication. The men who shared in the creation and development of this corporation are properly proud of this achievement. Competence in business and technical ability were mobilized to create an effective economic instrument, which served its area well. The record of this experience will have value for the audience interested in business and the technology of the energy field.

To the historian, however, this volume has a broader purpose. My professional career has emphasized the study of recent America, with particular attention to the industrial twentieth century. No one can work in this field without being impressed constantly with the pervasive but undefined influence of the great corporation, or without feeling the need for a more realistic understanding of its operation.

Such a study of the corporation system cannot be made from the outside. Subjective moral judgments, when they do not recognize or understand the alternatives actually available to the decision maker, are irresponsible and can be quite wrong. The significant questions involving public interest and social consequences transcend financial data and the published balance sheet, but are a legitimate field of inquiry. Unfortunately, research in the actual working of the nation's economic institutions has too often been stopped at the door of the business establishment.

The invitation issued to me to make this study of The Detroit Edison Company was accompanied by assurance that I would be given full access to documents, public or confidential, and that I would have full freedom in their use. For the confidence implied, and for the generous fulfillment of the promise, I am most grateful. It was an opportunity to use the tools of the historian to examine the working of this one of America's great corporations. I hope the resulting volume proves to be a contribution to the history of the United States, and thus to our understanding of America.

The leaders of the business community and the representatives of the general public do not agree, even within the respective groups, on what is the role of the corporation or on the distribution of its responsibilities. To the question implied, this work offers no general answer. This history has attempted to identify the meaning of the word "responsible" in Company usage, and to trace the changes and enlargements of the area of responsibility as they have been worked out and accepted in practice. They have grown with the Company.

At this point I should mention the literally scores of officers, employes and retired persons who found materials at my request, and who gave assistance in other ways. In the appropriate location, I have discussed briefly the source materials used. That section, entitled Acknowledgements and Sources, carries my expression of appreciation, and recognizes in a small degree the assistance I have received.

R.C.M.

THE FIRST HALF CENTURY
1903 - 1954

Introduction

Detroit is one of the older cities of the United States and celebrates 1701 as a birthdate. The tiny village and the unpretentious log fort were a bit of provincial France transferred to the spot overlooking the Detroit River, and the city is proud of its French origin. For almost one hundred years, the French flag, and then the British, marked their control.

In 1796, by agreement, the British soldiers withdrew, and Detroit began a second era. The advancing army of General Anthony Wayne established firm control, and the flag of the newly rising nation now flew over the fort. Detroit had become an American city.

A third era began in 1896 when the cough of the automobile first echoed through the quiet, tree lined streets of the old town. It was the birth of an industrial metropolis, or perhaps more accurately of an industrial system. Detroit would not again be either quiet or sedate.

In its early years, the Edison Company was closely linked with the development of the automotive industry—the two industries beginning almost simultaneously. Henry Ford, incidentally, was an Edison employe as he worked to perfect his first machine. From then on, the interconnection has been close, and The Detroit Edison Company has played a conspicious part in what happened in Detroit and its area.

The automotive industry and the city which is linked so closely with it have both had fantastic growth rates; The Detroit Edison Company over its first half century experienced a parallel growth rate. Those were years of great achievement, in which strong men hammered out the technology of an industry, and built a mature and competent corporate structure to go with it. The story of that period in the life of the Company is presented elsewhere in detail. The history of the Company was, in fact, commissioned as a part of the Company's celebration of its first fifty years, and it is not the purpose of this volume to retell that story.

However, events through World War II are reviewed here in

brief form, in order that the men and events of the more recent period will be understandable.

In the growth of that first half century, Detroit Edison was a fortunate company in a most favorable location. But Edison was not only a beneficiary, it was a contributor in a creative industrial process with which Detroit broke new paths. The history of the Company is thus a part of the history of industrial America and of the twentieth century.

Chapter One

BEGINNINGS IN THE INDUSTRY

Detroit at 1900 / The Historic Fruits of Competition / *Edision Illuminating* and Technology / The Company as a Corporation / The Corporate Structure / The City and Its Industries.

At the turn of the century, Detroit was one of a number of small but thriving cities located on the Great Lakes waterway. The population was in 1900 only 285,704. Chicago was much larger, but the other cities of the lakes region were close to the same size. The future urbanization of Detroit had been felt, however; the 1880 population had doubled by 1890, and doubled again by 1900.

It was not visible in the center of the city, however, where Detroit's many wealthy families lived. Detroit had been rebuilt, after the disastrous fire of 1805, after the model of Washington, D. C. The heart of the city was an open area, on which a number of streets converged, and was referred to both in casual conversation and on the maps of the day by a classic title, Campus Martius.

In the center of the Campus was the inevitable homely Soldiers and Sailors Monument. Detroit's new City Hall stood on one side of the open area, and the even newer and more ornate County Building was opposite it. Fronting the Campus on the north was Detroit's distinguished hotel, the Russell House, and opposite it was the then famous Detroit Opera House. Detroit's first skyscraper was there, and increasingly the newer merchant houses were moving to the Campus area, and even beyond.

South of the Campus, down Woodward Avenue, was the river a half mile distant; this was the older business district, but the better residential district was close to it. To the left, overlooking the river from a low bluff was Jefferson Avenue, lined with magnificent homes. To the right, west of Woodward, it was Fort Street which overlooked the river; the homes here were less palacial than on Jefferson, but they were distinguished. Woodward, beyond the Campus, was also a beautiful residential street.

The suburbs, to which the new clanging street cars gave access, housed the newer industries, and usually also those men who worked in them. It was not an affluent society, as that word would be used today, but it was stable and would serve Detroit well as opportunity developed.

The Detroit Edison Company, as presently incorporated, dates from 1903, but that date does not mark the beginning of electrical service in the city and state. In Detroit, as elsewhere in the world, the nineteenth century discovery or invention of the generator for the production of electricity (as opposed to the older battery) set local handymen to experimenting. The arc lamp is the simple development from the electric spark, and everywhere companies attempted to turn this newest scientific discovery to commercial advantage. In Detroit, there were several such companies, using different patents and with differing electrical characteristics, but all of them offering arc light service. The arc lamp was a spitting, hissing, intensely brilliant source of illumination which was magnificent on the street, but was tolerable inside only under special and unusual circumstances.[1]

Despite the patent fights and the rival claims, the arc companies were substantially of one technology. The lamps were wired in series, like the old-fashioned string of Christmas lights. The substantial drop in voltage at each lamp required that the initial voltage at the machine be quite high. This allowed distribution over considerable distances, but the high tension lines, inadequately safeguarded, were a menace to public safety.

It was such a system as this which allowed a local company to bid, at last successfully (1884), for the privilege of illuminating

Campus Martius, about 1900. The Light towers are in front of the then new city hall.

Detroit's streets. The protests of the displaced gas company were public and violent. The successful company, using Brush patents and the Brush name, brought its characteristic system of lamp placement. In Cleveland, where they had an early contract, but most extensively in Detroit, they installed a lamp or a cluster of as many as four lights on top of a tower a hundred to two hundred feet tall. The effect seen from a distance was spectacular, though in the beautiful heavily wooded residential streets of Detroit, the effectiveness was questionable. For many years, however, even after the city's own municipal plant took over the lighting of the streets, the tower system continued.

With two and sometimes more companies bidding for the city's street lighting contract, the competition became severe, and ruthless. The successful bidder might well find that he had cut his price to an unprofitable level, while the unsuccessful competitor could pick up only fragments of private business. The battles over the contract were dramatic, bitter, public brawls which inspired rumors of excessive profits and political corruption.

Detroit's decision to build and maintain the Public Lighting Commission's plant (1893), as an alternative to contracting with an arc light company was well nigh fatal to the two surviving arc companies, for there was hardly enough private arc business for one. The result was a bitter and destructive competition for this surviving business.[2]

Quite removed from the embattled arcs, and pursuing its undisturbed and quite profitable way, was the one Detroit company entitled to use the famous Edison patents. The little glass globes of the Edison Illuminating Company, which were the unique mark of its service, would have been as inappropriate on the city's light poles as the arc lamp was inappropriate inside a store or residence.

Thomas A. Edison had nothing to do personally with the Edison Illuminating Company of Detroit; the name merely proclaimed the fact that this company had the exclusive local rights to use the highly publicized incadescent light system which Edison had invented. The license was issued by a New York corporation bearing the name Edison Electric Light Company. By 1886 Edison himself had almost nothing to do with this company or its management either, but there too, the name was considered an important busi-

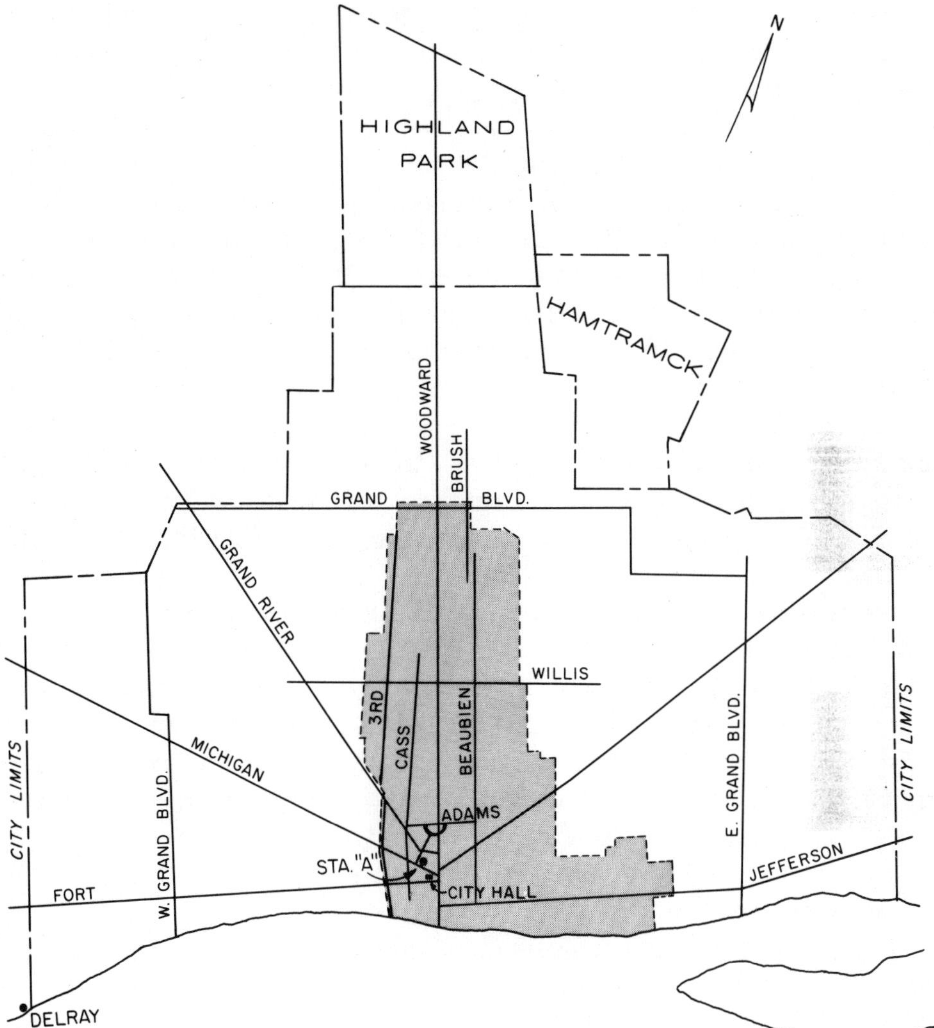

GREY AREA INDICATES ORIGINAL 'EDISON ILLUMINATING' CO. — CIRCA 1900

ness asset and the control of the patent right was the essence of the license.

The Edison Illuminating Company of Detroit was locally organized and financed. With the advice and technical assistance of the New York organization, the Company built in 1886 the long familiar Edison building standing on Washington Boulevard. The building created a good deal of excitement in the electrical world, for it was said at the time to be the largest power station ever built for an electric company.[3] Boilers and engines occupied the major part of the first floor. These engines, with a pulley wheel at each end of the drive shaft, were belted to two Edison generators on the second floor. To the contemporary industry, this building was a marvel of compactness and efficiency.

The system used was Edison's 110 volt direct current. The advantage was great, for the factor of safety was important, and even more important was the availability of the Edison lights—or burners as Edison continued to call them—which could be lighted and extinguished singly as easily as could a gas jet.

The great disadvantage of the Edison system was the fact that the low voltage at which the system must operate limited the area which could be served to roughly a one mile circle about the power plant. Within that circle, however, the little sixteen candle power carbon lights became a kind of mark of elegance, the quite conscious indication that the householder could afford the finer things.

In the closing years of the nineteenth century, then, there were four plants or systems operating in this small city. The city plant, with capable management under the direction of the Public Lighting Commission, provided the electric current for the towers on the city's streets. The two arc companies, exhausted by their competitive struggles, tried to work out an uneasy and imperfect merger agreement. The fourth, the strong and well-led Edison Illuminating Company, sought, in its own defense, an end to the chaotic state of the industry in Detroit.

On the national scene, a series of mergers had brought together the major competitors in electric equipment, and allowed a rationalization of the patents. On the local scene, the old Brush company, currently owned by Peninsular, was purchased by some men chiefly interested in Edison Illuminating, and then was offered to that Company on a lease-purchase agreement. The other company fell directly to Edison, whose stock was exchanged for the company's bonds in a negotiated transfer.

The Detroit Edison Company was not created to unite or consolidate the electrical services of Detroit. This had been done well before the birth of the new company by the failure or outright bankruptcy of all but one of the older private companies.

The Company was made necessary by events which followed

Station A, the Washington Boulevard original plant of the Edison Illuminating Company of Detroit.

this consolidation. The Illuminating Company was heir to at least four power plants, incompatible in voltage, frequency and operating equipment, and containing some machines so old and so inefficient that they were, in fact, junk. Temporary improvisation could meet the immediate needs, but the consolidation forced some men in the industry to think about the present and prospective needs of total Detroit.

The purpose of The Detroit Edison Company, as a corporation, was to finance and build a new power house to supply these needs. At first, the proposal was that the new company be created just to own the producing facilities, with power to be sold, for distribution, to the independent companies. Almost at once, however, the decision was made to purchase these distributing companies.

This was an enormous proposal, far beyond anything Detroiters had experienced. Detroit was either unable or unwilling to finance the undertaking so the opportunity went to a New York holding company, The North American Company, which had both experience in, and access to, finance. That company organized the syndicate which did the financing, secured the incorporation, set The Detroit Edison Company on its feet, and retained representation on the board of the Company. Detroit owners of stock in The Edison Illuminating Company were offered an opportunity to invest on favorable terms, but only a portion of them did so.

New York was the state of incorporation, in part for the convenience of several of the men involved, but chiefly because the law of Michigan then set impractical limitations on the fiscal size of such corporations. Since Michigan also prohibited a foreign corporation from doing business as an electric utility in the state, a measure of ingenuity was required in shaping the corporate form.

To meet the requirements of the state's law, and perhaps also in deference to the antitrust mood which already marked the Progressive era, The Edison Illuminating Company was maintained. The Peninsular Company was released from the purchase agreement and reconstituted as a wholly owned subsidiary. For a number of years, customers' bills continued to be made out by these companies, despite the fact that Edison in reality totally owned them.

Two developments in technology made this new business venture possible. Critically important was the perfection of the transformer. This simple device allowed the efficient shift of an electric alternating current to high voltages for economical transmission, and the return to low voltage for safe local distribution. It freed the industry from the limitation of the one mile circle which had characterized the Edison direct current system. With this device, the new plant at Delray could produce and transmit power at 4,600 volts to an area as large as the county and even beyond.

The second technological development of significance was the adoption of the turbine instead of the reciprocating engine. Even in this primitive form, the first 3,000 kw machine at Delray was a marvel in size and in efficiency, instantly making obsolete every other machine in the Company—and indeed in the whole area. It was this efficiency of central power, and the economy of relatively high voltage transmission, which made Detroit Edison service attractive.[4]

The only significant change in the corporate structure, as it was established in 1903, occurred in 1916, and was the result of a series of preliminary actions looking to this end. The full significance of this decisive change was probably not appreciated until, in later years, the word "holding company" acquired a special and unfortunate association.

All about the central city there were areas which had no electrical service, and occasionally some villages which had local private or public plants now technically obsolete. In the years following 1903, the service area of the Company expanded in constantly widening half circles, as facilities and finance became available.

The old interior part of the city remained a direct current service area under The Edison Illuminating Company. The balance of Wayne County was served by the Peninsular Company using the alternating current for those greater distances. After a good deal of improvising, the region to the north and generally beyond Wayne County was united into the Eastern Michigan Edison Company, which became the third totally owned distribution company. The Essex County Light and Power Company (1912) was the result of an ill-starred venture across the river in Ontario.

Historically, when the Company had purchased a village plant either from its private owners or from the municipality, it had inherited the franchise and other legal privileges or responsibilities. Sometimes local sensibilities, or the law, made it convenient to maintain the structure of the acquired company. The net result of all this was that Edison as a holding company owned several other holding companies, some with subordinate structures even to the fourth and fifth strata.

There was no abuse here of the sort which later brought discredit to the name of holding company. It would have been quite easy and proper to let the inherited structure stand. Instead, in a series of moves, ending in 1916, the whole structure was simplified. The subordinate companies were either eliminated or were made instruments of convenience only. All property, all obligations, and all responsibility were transferred directly to The Detroit Edison Company. It was a simplification which by intent made the business behavior of the Company thereafter transparently clear.

From 1916, or as soon thereafter as it could be achieved, the Company served a single unitary service area with a single uniform rate schedule. The operating properties were brought under a single mortgage, which guaranteed a single series of bond issues, and a single type of stock represented the equity interest.

The transfer of the regulatory authority to the instrumentality of the state (1921) tended to allow the rules and decisions of the State Public Service Commission to supersede local or municipal requirements, and thus confirmed the application of a single policy to the whole territory.

As a very young corporation, it was Detroit Edison's good luck to serve the most fantastic growth industry of all time. But even before the automobile, Detroit had a sound manufacturing tradition.

In the outskirts of the city there were metal workers manufacturing stoves, others casting wheels or shaping other metal parts for railroad cars. Along the water front and in the adjacent streets there was a complex of machine shops and small plants engaged in the manufacture and repair of marine engines and gear for the extensive ship building and shipping industry. Tradition has it that the metal working industry was especially dependent on the original German immigrants to the region, and on the successors who learned the trades from them. Certainly the level of skill in metal working was high.

It was to these shops and these craftsmen that the pioneer auto

In this Washington Boulevard plant of The Edison Illuminating Company of Detroit, Henry Ford (r) was chief engineer (1803-4).

men took their problems; these unrelated establishments turned out parts, sometimes almost hand made with blacksmith tools. The first automobile firms did hardly more than assemble the manufactured parts.

Detroit had the skill, and it also had daring. The late nineteenth century commercial prosperity of the city, combined with the vigorously exploited lumber industry, provided the well named 'venture capital' for the new industry. According to one local historian, there was a time before Ford moved from central Detroit to suburban Highland Park when half the automobiles produced in the world were being assembled within a mile of the present Wayne State University main campus. There were a good many men who ventured to invest time and money in new enterprises, and a few were fantastically successful. Detroit could not escape the excitement of such adventures, or of such rewards, or shake the feeling that this was her destiny.

The ultimate impact of the industry, however, was much more significant. The Ford Motor Company was in many ways the first exponent of an industrial method which would revolutionize the world. It is possible to isolate three elements in this method. It is possible also to argue that Ford did not originate any one of them, but this is irrelevent. Henry Ford and his associates put them together, as Thomas A. Edison had put together the elements of his electric system, and from the parts created a totality which was new.

1. The multiple production of identical parts. This is elemental to any mass production operation, but calls for such quantity production as would justify highly specialized machinery of the sort promptly made semi-automatic.

2. The moving assembly line. This was not new, though it was not common. It meant work carried to the worker, with tasks so simplified and repetitive that anyone could learn any job by doing it.

3. The timed integration of the whole factory. This was the key to the operation, and its chief novelty. From individual special purpose machines and from sub-assembly lines came streams of parts and manufactured units, with men and machines so exactly organized and so precisely timed that the whole functioned as if it were one intermeshed unit. The factory had itself become a

machine, and the architect of the building in Highland Park was asked merely to enclose a process.

It is the process which is the exciting thing. It multiplied the fruits of man's labor and, in so doing, gave to him a production instrument never dreamed of. Ford's product changed the nation's transportation, his five dollar day altered the goals of labor more used to ten dollars a week, and his millions made a myth of the man in his own life-time—but these are at best of only relative significance. Detroit came to mean not just the quantities of automobiles produced, but the revolutionary method of their production. Applied first to this industry, and then wholesale to others, it produced plenty for days of peace and strength for days of war.[5]

What this did to the old Detroit of tree-lined streets and quiet charm can be viewed with shocked regret or enthusiastic approval as interests dictated. In 1907, when Detroit Edison was new but well established and before the automobile was a major factor, the Company estimated its area's population at one half million. By 1917, the number was doubled to one million, and doubled again to two million by 1926. Depression, war, and other massive disturbances slowed up the process of growth, but the year 1947 counted three million people in the vast industrial area in and about the city. A multiplication in population to six times in four decades is not a record, but it has seldom been matched in so large a city.

The Company's involvement in all this is best given in statistical summary. The thousands who flocked to Detroit in pursuit of Ford's Five Dollar Day—and their successors—became domestic customers, and Edison profited. But Edison, almost annually from 1915 onward, poured half its vastly increasing kilowatt production into the machines and production lines of the region's industry. Electric power, so flexible and above all so available, was the ideal complement to the new factory method.

CHAPTER TWO

THE APPEARANCE OF CRISIS 1939 - 1945

The Forces of Disorganization / The Pattern of Leadership / Policies and the Personal Tradition / Capital and the Fixed Dividend / Labor and the Paternal Tradition / The Customer and the Rate Case / Law and the End of an Era.

A COMPANY IS NOT MERELY LINES AND MACHINES, AND ITS REAL STRENGTH is not defined in law or the Articles of Incorporation. The company is also men and the methods which they employ. The written rules and procedures are, of course, important; but of even greater significance is that mass of practice, custom, attitude, and mood which interprets the rule and molds it to the specific case. The corporation, save in a narrow legal sense, need not be impersonal or mysterious and is not self-directing. The Detroit Edison Company had long been an instrument through which the attitudes, hopes and resolves of its leadership found their expression. The Company of 1939 can be explained best through the men who made it and who continued to guide its destiny; Alex Dow, the veteran president, had given the organization the stamp of his personality.

This association of the Company as such with the men of management was an asset, for the men were strong. It also created a concern for the Company when that leadership was jeopardized, for the Company itself appeared to be involved. To say that Alex Dow and The Detroit Edison Company were one was in an earlier day no more than a rhetorical exaggeration; to say it when Dow approached eighty years of age was to imply an impending and inevitable crisis.

The appearance of crisis came not only from those changes which time itself inevitably brought to Edison and its aging leadership. It came also from the discovery that some policies and practices hallowed by time and association were no longer relevant to the world in which Edison must live.

There were three separate but distinct forces which, singly and jointly, imposed upon the Company the necessity of adjustment, or at least of reconsideration before decision. They were forces, too, which would surely continue to exert pressure into the postwar world.

1. The Holding Company Act of 1935 at first glance appeared to be of minor importance to the Company. By 1939, however, its provisions were held to be within the constitutional power of Congress, and specifically to relate to the Company in several important regards. The extent of the disruption of leadership and policy could not at that time be estimated, but certainly the board itself would be in some degree changed.

2. The ultimate import of labor legislation, including the Labor Relations Act, also was not clear. The Company had been spared direct involvement in the violent labor conflict of the thirties, but wartime regulations, labor shortages, and legislation combined to indicate that labor decisions were no longer to be casual or incidental to the Company.

3. Unfortunately, the war-born distortions of costs, prices and profits caused the Company to be charged with profiteering. The charge resulted from an odd quirk in the law, and from its formula for establishing the "normal". For the moment, the controversy provoked was quite serious. It would become more so if it interfered with efforts necessary to anticipate and provide for the electrical demand of economic Detroit.

Three people in the Company had impressed their personalities on aspects of management practice. It had been a tightly knit leadership, perfectly balanced and consistent, and presenting always a united front to the world outside.

Alex Dow was the venerable president of Detroit Edison. He may have been given the more august name of Alexander at his christening, but if so no one remembered it. He was a Scot by birth, who had come to the United States at about the age of twenty. He came from a labor class family, and in his native land his chances for formal education had been sharply limited. He landed at Baltimore with some practical knowledge of fires and furnaces, and little theoretical training in electricity. Perhaps it was this interest that led him to seek jobs where electricity was involved and, like many of his day, he learned by doing. In a surprisingly short time, he was in charge of the arc system which lighted South Park, Chicago.

A recommendation led to a job interview in Chicago, which resulted in his appointment (1893) as manager of the municipally owned light plant being created by the Public Lighting Commission of the City of Detroit. He was chosen over a local man with experience in electrical work—Henry Ford. At this stage in his career, Dow was thirty-one years old.

He moved from this post in 1896 to another Detroit assignment as manager of The Edison Illuminating Company, and he kept that title when that company became the base of the new Detroit Edison in 1903. Since the elected president was in New York in those first years, Dow, as vice president and manager, was in charge of those matters which concerned Detroit. His election to the position of president in 1912 only confirmed what had been fact.

By 1939 Dow had been in control forty-three years. He had outlived his contemporaries, and no one was about who could remember when he had not been in command. His was a long and distinguished career, filled with public service. He had grown with the industry, and stood as its acknowledged dean. The handicap of inadequate education had long since been overcome, and he had numerous honorary degrees and citations. The Company had grown strong in his confident hands, and he was frankly proud of it and the reputation it enjoyed in the city, state, and nation.

The second member of the management group was Alfred C. Marshall. He was younger than Dow, a gifted engineer who had worked first under Dow at the municipal plant, and who had come to the Company in 1908. He rose rapidly in the Company, in time was named vice president, and was substantially in charge of the system. In 1923, he was designated general manager also, and it was clear that Dow intended him as his successor. In 1923, he was just past fifty years of age.

Fifteen years later when Dow announced his retirement, Marshall himself was close to retirement age, but at long last he was president. Dow, however, retained his office suite, his position as chairman of the executive committee of the board, and his habit of command.

An able engineer, Marshall lacked the flair for publicity which was so characteristic of Dow. Slender, dignified, always immaculate, with striking white hair, Marshall gave the impression of gentleness rather than strength. He had been so long under the shadow of the older man that he was less well known than his achievements merited.

Dow talked a great deal about customer relations, and his theoretical statement about Company posture was in every way admirable. But, when he had to deal with a customer himself, he was likely to show impatience when he encountered stupidity in matters obvious to him, and his volatile temper was well known when he suspected fraud or the intent to deceive. Fortunately, this whole problem of customer relations he could and did move over to an extraordinary woman, Miss Sarah Sheridan. She was the third member of the "management" team, and became a vice president in 1941.

Miss Sheridan was a tradition maker in the Company second only to Dow himself. She had been a clerk in the office of one of the arc companies, and came to Detroit Edison, Dow was accustomed to say, as the most valuable asset with that acquisition. She was a woman of great charm, to whom tact was probably a natural virtue, but who contrived ways to make it contribute to the Company. Her first responsibility had been merely to collect bills, but in time it grew to include all of sales, service complaints, and indeed all aspects of customer relations. Again as Dow liked to say, the Company had two parts, production and sales, and Miss Sheridan had charge of sales. This statement was not quite accurate, for it left out whole aspects of company responsibility, but it did emphasize the wide and important role Miss Sheridan held in the Company.

Alex Dow and his successor as president of The Detroit Edison Company, Alfred C. Marshall.

Sarah Sheridan, from a portrait by John Coppin, owned by the Company.

To complaining customers she was sympathetic and understanding. Her dignity and her consideration were totally disarming, and she had the authority to make good on promises. To her, and to the president with whom she worked so closely, the prime consideration was retaining customer goodwill.

It was a gifted trio which for nearly four decades managed The Detroit Edison Company. It was a singularly harmonious group tied closely by links which were both professional and personal. Beyond question, Dow was the leader, the ultimate authority. The other two were strong, each in his own area, because they had won Dow's confidence, and because neither could have considered opposing his decisions. Such a harmonious and effective group of leaders is rare to any company, and the high success of The Detroit Edison Company was in large measure because of that direction.

It is not surprising that this leadership exhibited in a conspicuous degree unusual and somewhat old fashioned virtues of duty, responsibility and authority. The ideas which Dow and his associates exemplified became impressed upon the Company. Whether in written order or merely in practice and precedent, these highly personalized traditions were as much a part of the Company as were the plants and lines.

The Company itself was an entity quite removed from machines and investments. Dow used the words "my company" when his share of the stock was slight indeed; the words carried no more the sense of proprietorship than did the words "my wife" or "my country." They were a symbol of dedication to an institution. Dow gave loyalty himself, and he expected it from associates and employes.

Dow often cited the principles by which he proposed to be guided. His first duty was to see to it that the Company was maintained in first-class condition. Beyond that, his only obligation was to do justice to the three parties of interest in Detroit Edison. The wording and the order might differ in his usage, but the three parties were always most clear: the shareholder, the employe, and the customer. The wording differed, and the source to which he attributed this set of principles changed widely from time to time. But the repetitious emphasis upon them, whether as the text for an address or to rationalize a decision, made them important.

In this statement of obligation, the word "justice" is at once the ideal and the problem. In the time of Dow's leadership, justice had an amazingly subjective quality. The fact of Company autonomy, and Dow's unquestioned leadership in it combined to create a responsibility almost completely personalized. Detroit Edison would give justice, and justice was what Dow said it was.

The simple three point obligation, and the highly subjective definition of responsibility combined to create a unique business philosophy. To the older generation, the principles and their application were self-evident. In the half decade which followed 1939, these practices of a life time were challenged.

Dow brought attitudes from his native land which he could no more lose than he could the soft burr of his speech. He had been much affected by the nonrevolutionary radicalism of his day in Britain, a radicalism which especially involved scientific circles. In Britain speculative land holding was evil. Here British radicalism touched hands with the American Henry George, who was influential on both sides of the Atlantic. Gain which came from mere speculative land holding, or for that matter from any kind of speculation, was illegitimate because it was not earned.

The vested interests on the continent, monopolizing land and privilege, had been attacked in futile political revolution. The effective remedy would come rather in the new world which science was in the process of creating—if only the scientific achievements were not themselves reduced to a vested interest.

This was in no sense an attack on capitalism, for the earning or creative power of money invested was recognized. Money when actually so invested, should receive a return both certain and fixed. Money earned its wages, but that was all. In theory, the stockholders were the owners of the Company, and both the president and the board served at their pleasure—but that was theory only. In practice, they were divorced from any meaningful participation in management or decision-making, and that included setting a dividend rate.

The dividend was established at a "just" figure, and was kept there. In 1911, the dividend was raised to seven percent and in 1916 to eight percent. The depression disturbed this practice (1932–35), but in 1936 the dividend was stabilized at six percent and that figure remained through World War II. The fixed dividend as an interpretation of "justice" was a firm part of the Company tradition.[1]

By 1940, however, the theory of the fixed dividend, the "wages" for capital invested, had been made obsolete by the active changes in the economy during the most recent two decades. There was much truth in the argument that the current policy did not, in fact, properly represent equity holdings of considerable magnitude, and that to be just the traditional policy of the Company must be changed. The matter became critical after the Holding Company Act of 1935 with its "death sentence" provisions.

The shadow of the holding companies and their impending liquidation disturbed the market for Company securities and made any new financing of importance almost impossible. The continued dependence on internally generated funds would be critically limiting if the postwar industrial activity hoped for in Detroit developed. Traditional policies seemed certain to create a crisis in finance.

Justice to the employe was equally subjective. The words "the going wages or a little better" were hardly definitive. In theory, all employes could appeal to Dow, but as the Company got larger and he was more and more removed from actual contact with the men in the field, the validity of this kind of personal justice could be questioned.

Dow himself came from a working class background. Perhaps it was this that kept him from becoming the "good employer" in the late nineteenth century's unhappy meaning of that word. He did not intrude on the private lives of his men. He demanded only a reasonable day's work from the crew which manned the powerhouses and built the lines—and beyond that each man was his own master. He defended and protected his men, and he won their loyalty.

In theory, the wage contract terminated each Saturday night. No rights and no obligations survived that pay day, and a new contract began each week. In fact, the Company granted sick leave, financial assistance when needed, retirement stipends, and in many other ways defied that theoretical wage contract definition. This was, however, a unilateral decision, a beneficence and not a right. As an employer, Dow would do justice to his men, but Dow would himself define that justice. There were no rules which would interfere with Dow's freedom of action.

Dow's conception of his role, or perhaps it was his Scotch caution, caused him to insert in such grants precautionary clauses which would prevent them from becoming enforceable rights. In large measure, the early labor struggles in the Company were an effort to secure just this kind of contractual obligation.

The long tradition of Detroit was that of the nonunion open shop, with the automobile companies setting the pattern. When, in the days of the depression, these plants erupted with labor wars and sit-ins, it was inevitable that the management of Detroit Edison should be attentive and concerned. The Labor Relations Act and the later provisions of the wartime manpower administration gave legal support and defense to the organizing efforts.

In the labor tradition of the Company, management had always been at pains to take care of its own, and it expected understanding and acceptance. The first employe vote which chose union representation was a shock; management could hardly believe that employes would choose "such leadership" in preference to that in the Company, or that confidence in the Company good-will was at so low an ebb. Marshall, and his subordinate and successor, James Parker, found the situation most disturbing.

Benevolence, no matter how well meant, was not enough. Justice, bestowed as a gift, was not justice. Gratitude is not a satisfactory base for a labor policy. These generalizations came easily to hand, and few would have disagreed with them as an abstract principle even in 1939. In practice they called for a new labor policy, the details of which were yet to be worked out.

The Dow formula was deceptively simple: to the worker, the going wage or a little better; to the dollar invested, a fixed and certain return, and no more. The surprise was in the final and significant third part: all the remainder was to return to the customer in lowered costs and improved service. All society was to benefit directly from scientific and technical progress.

It was a surprising formula, though by constant repetition it became so familiar that sometimes its more extreme implications were forgotten. It was a creed which made both Dow and the Company he led more than a little hard to understand by other business men.

The guide lines were quite clear to the Company's leaders. In the conventional economic society, price was determined by the competitive process. Dow had watched the struggle between the arc companies in Detroit, and drew the moral therefrom. Competition will not work in the electrical industry. Duplication of lines and facilities is a dangerous waste of capital, to no good purpose. Price, since it cannot be set in this industry by direct competition, must be an administered price, which means substantially a monopoly price. Dow did not shrink from the next logical conclusion: an administered price then must be set with full public knowledge, and must rest on public approval and confidence. Public participation in the price-making process, either directly or through a responsible public agency, was an essential corollary to the industry's inevitable monopoly position.

No legal approval, however, could give validity to the actions of a suspect and unpopular public utility. Public approval and public confidence in its integrity were a utility's best protection. The Company must convince the public that its best interests were being served, and the way to do this was to serve the public interest. In this, Sarah Sheridan was invaluable. She succeeded because she—and Dow—honestly felt that the Company interest and public interest were totally compatible, or were indeed indentical.

For many years, improving technology and increasing volume allowed Edison on occasion the luxury of price reduction. For a number of years beyond the legal requirement, Dow went through the formality of informing the Detroit city council of the rate structure, ostensibly to get their approval. Actually, he enjoyed these occasions and the response he received when announcing a price reduction.

In the difficult days immediately following World War I, prices and wages both demonstrated an erratic rise. It was the end—though not everybody recognized it—of the older fixed or stabilized economy, and the beginning of the extreme swings to which the decades since have been subjected. In June, 1920, the Company made the first substantial rate increase in its history.

Unfortunately, the reception of this increase became entangled in the political struggle over the ownership of the Detroit street car system. The end result was the appeal from the city to the new State Utility Commission, and the acceptance of jurisdiction by the state. In view of the size and complexity of the Company which extended far beyond the city's limits, the purported city regulation was already an anachronism. Dow and the Michigan legal counsel, James V. Oxtoby, pushed and prodded the State Commission to action, and attempted to use its hearings and proceedings as a means of informing the customer public and the state at large.

Almost the same kind of difficulty appeared in World War II, and the Company found itself, in spite of all its efforts, being denounced in the press and from the politician's platform. The facts were quite simple. The high load factor when war plants ran around the clock, combined with reduced expenses from delayed maintenance, reduced employment, and postponed extensions, all combined to increase the profit margin markedly over the prewar average.

The Company was aware that this was a wartime phenomenon and was accordingly reluctant to see a permanent change in price structure. It was a situation which management found most uncomfortable. All the talk about public responsibility and the compatibility of interest of Company and customer seemed to be contradicted by the hard phrase "excess profits."

There was nothing wrong with the tradition (or with the rates themselves) and when times returned to normal, as they would, the rates would again be fair. But in the meantime, for reasons which were totally external, the Company faced a challenge and it was in no position to defend itself, least of all by an appeal to its own tradition.

To a Company so nearly autonomous, so proudly self-directing and so accustomed to public order and acceptance, the clamor from stockholders, from employes, from customers, and the public seemed bewidering and dangerous. The men in the Company had appealed often to the ideal of justice, to be administered to shareholder, to employe, and to the public as customer. In each of these areas, the best of intention had not prevented bitter disagreements at times.

From the time the Holding Company Act of 1935 was upheld in the courts, the Company was in a tangle of hearings and court cases involving jurisdiction and found itself forced to actions which recognized Federal jurisdiction.

The effort to reestablish independence led in 1939 to a long, difficult and sometimes insulting hearing at the hands of the Securities and Exchange Commission examiners. The issue was exactly this one of independence with the implication of wrong doing in every question. Dow found himself under oath, and giving answers to questions involving events long since forgotten, or episodes and documents presented out of context.[2]

Dow, in spite of his unusual vitality, knew that his time had come. At the first meeting of the board following the SEC hearings, he offered his resignation. He was seventy-eight years old when he retired in 1940, and eighty when he died in 1942. In 1943, both Marshall and Sarah Sheridan independently offered their resignations, to take effect the next year. With the loss of Dow, Sheridan and Marshall, the brillant team which had so long typified Detroit Edison, was gone.

This was not so much a crisis as it was the end of an era. The policies which these people had developed were good policies, and the implementation of them had been the basis of a proud career in the Company. But if there was anything clear in the closing days of the Dow-Marshall regime, it was that the situation was changed. The traditional policies had become formulae. The principles of justice, equity, responsibility would still rule but policy must be altered and restated in more modern style under a new leadership.

CHAPTER THREE

THE BOARD AND ITS CHAIRMAN 1944 - 1954

Federal Control and the United Company / The Responsibilities of the Board / Prentiss Brown as Chairman / Finance and the Holding Company Involvement / The Solution of the Rate Case / Congress and the Canadian Interconnections / Brown as a Public Participant.

THE HOLDING COMPANY ACT OF 1935 IS A LANDMARK IN ANY HISTORY of an American public utility. The holding company was an instrument relatively old in American corporate history, and one which was frequently useful. Unhappily, especially in the decade of the twenties, the name became associated with financial practices which were risky and sometimes fraudulent. The guilty and the innocent were then lumped together in the press and were condemned in law. The Holding Company Act of 1935 was intentionally punitive, and represents that mood at its New Deal extreme.

Because it was a utility and because of its historical relationship with The North American Company, the Company had voiced its opposition to the law prior to its enactment. It is probable that the subsidized campaign of utilities and holding companies added to public suspicion and increased the likelihood of its enactment. Once it passed Congress, the law began that long and involved journey through the courts to establish and to confirm its validity.

The first reaction of the Company to the enactment was a hurried examination of its subsidiaries, lest it be called a holding company itself, and a careful checking of distribution lines and contracts to be sure no power crossed a state line and thus established Federal jurisdiction in interstate commerce. Neither of these actions proved important.

As to the Company itself being "held," the suggestion seemed unthinkable and unsupportable to the officers. This was the contention despite the fact that two of the big holding companies each owned above twenty percent of the Company's stock. When, in 1937, these two companies registered under the Act, The Detroit Edison Company asked for a hearing and an opportunity to prove its substantial independence.

It was these hearings (1939) which so disturbed and troubled the older Edison leadership. No abuse of the law was disclosed and no scandals unearthed, though by suggestion and inference some personal friendships and business relationships of long standing were made suspect. The Company was outraged at the insinuations and even at the need to defend itself, but there was no denying the fact that, by the rigid definition of the law, it was a held company.

The holding company which had created Edison was ordered to divest itself of its interest; the resolution of the problem with The North American Company was relatively easy. That corporation proposed and carried into effect over several years a series of stock dividends by which it transferred the Edison stock directly to the individual shareholders of North American. To aid North American in this distribution of shares, the stock of Detroit Edison was changed from par of one hundred dollars to par twenty dollars, and exchanged five to one for the old shares. This change occurred in 1941. When the shares of Edison's stock which remained in North American hands were reduced to two percent of those outstanding (1942) the SEC order was deemed satisfied.

The American Light and Traction Company which also held approximately twenty percent of Edison stock could not use the simple device of the stock dividend. That company was itself held by The United Light and Railways Company and a mere transfer of the stock by dividend to a parent company would not accomplish the purpose of the order. The United group agreed that the stock received in the liquidation would be offered for sale. With this arrangement the United group became vitally concerned with the price of the stock in the market; the market high in 1944 was less than one-third the high of 1929, and was well below even the depression lows of 1935–6–7.

The United group insisted that its interests differed not at all from those of the other stock holders, and several disinterested members of Edison's board agreed with them. They could point out that in the fifteen years since 1930, there had been no increase in stock with the exception of the unimportant split already mentioned. The long-term debt actually was no greater in 1944 than it had been in 1932, although the plant, after depreciation, was listed at seventy-five million dollars increase. Neither the posture of the Company nor the market price of its stock reflected the reality of its financial position.

The United group hoped in some fashion to modify that. They could not hope to recover the predepression price they had paid for the stock, but they could hope to recover at least a more reasonable return on value even in this forced sale. They demanded, and after a rather bitter dispute received, two seats on the board of directors (1944). Two indignant Detroiters, James S. Holden and Wesson Seyburn, angrily resigned from the board on the decision to seat the United representatives, and thus the vacancies were created for them.

The men who came to the board of directors were themselves quite acceptable and they or their successors continued to sit for as long as the United Company held its stock interest (1949). Their presence, however, symbolized an undercurrent of concern with the Company. They demanded what they called a "strong leadership" and the security and confidence which would be reflected in the improved price of bond and stock.

In any corporation, the legal authority for policy and personnel is vested in the board of directors which, since it has ultimate authority, is presumed to have ultimate responsibility. It is an assumption which is often found to be in error. Delegation of authority can become habitual and, in an organization sufficiently complex and technical, the board can become almost captive of a corporation's management.

In some degree, this had been the story of Detroit Edison's board. Confidence in the long-time management was expressed in an almost automatic approval of recommendations in men and policy. The stockholders themselves had even less to do with policy or management. Their meeting was held in New York, where the management's slate for directors was almost never challenged even by a counter nomination; there had never been a proxy fight, and only one threat of one marred the corporate record.

By a kind of unwritten rule, the board had come to represent those three interests which Dow so often mentioned as his responsibilities. The North American Company owned or controlled a number of utilities but, for reasons chiefly historic, chose to limit its continuing participation in Detroit Edison. The actual extent of its holdings varied from time to time. The North American Company had designated one or more of its officers to represent that interest on the board.

The important legal firm, Sullivan & Cromwell, was the New York general counsel for the Company as well as for North American, and a partner of the firm by tradition sat on the board. The problems of marketing of stocks and bonds, and of other financial transactions usually, but not always, suggested the desirability of a member who would provide direct access to New York's money markets. In a nine man board, this New York representation was normally three members.

Technical and industrial skills were present on the board, at first by appointment from the outside, but with mounting confidence the veteran officers of the Company were so recognized. This representation of the Company officers, its employes, and its staff tended also to be about one-third.

More and more as time went on, the identification of interest of the Company with the city itself was recognized by the election of board members who could be thought of as customers and representatives of the public in general, i.e. the public of the Edison service area.

These proportions were not set down in rule and were, on occasion, allowed to vary when circumstances made it desirable. The atmosphere of the board meeting was relaxed, for the members seldom took recourse to parliamentary rule or formal vote. It was a prestigious body, on which able men served with pleasure and satisfaction.

Such a board, traditionally strong both in resource and in experience, had given the administration a sense of confidence and assurance. In 1940, of the nine man board, two had held office since the incorporation in 1903 and six members boasted fourteen or more years of experience. The board gave every evidence of both strength and stability.

Unfortunately, the board itself was altered drastically in the years 1939–44. The changes in board membership came from causes inside and outside the Company, were both personal and legal, and time itself was a contributing factor. The war forced two men to leave the board. Two resigned over an issue in policy which developed when the United Group gained representation. The impact of the Holding Company Act of 1935, as it was applied to fiscal institutions, forced other resignations.

Including the death of Alex Dow, the years 1940–44, inclusive, saw nine vacancies in a nine man board. Marshall had been a director for twenty years before his retirement from the presidency in 1944 and kept his directorship in retirement. Two New Yorkers had held office for ten and twenty-one years respectively. Of the other six directors, the longest tenure was since 1940.[1]

The sudden and drastic alteration in the membership of the board came at a moment particularly unfortunate for the Company. New men must choose the new administration. The war and the critical postwar years demanded decisions which challenged precedent and established policy. Finally, for the first time, the Company found itself operating within the unfamiliar confines of a Federal regulation, for the Courts on appeal in 1941 held that the Company was indeed directly subject to the jurisdiction established in the Holding Company Act of 1935.

There was neither panic nor a sense of crisis among the young men who had the work to do. Over the years, A. C. Marshall had shown a particular capacity to recognize and develop able young men in the organization and now, in staff positions, they had confidence in each other and in the Company. The wartime achievements, both recent and still current, justified that faith.

At the moment, however, the problems which loomed were external to the Company itself, and often involved relationships in various levels of government. It was in these fields that the men in the Company had had less experience; here, it could be argued, the Company needed outside assistance.

Prentiss Brown was offered and accepted a newly created post inside Detroit Edison. His title was that of chairman of the board, and he was designated as the chief executive officer. Brown was a native of Michigan, a product of Albion College and of the University of Illinois, and a life-long resident of the Upper Peninsula. A career in law led into politics and to the House of Representatives and then to the Senate of the United States. Defeated for reelection, he accepted appointment by Roosevelt to head the Office of Price Administration. This was a difficult job, and an unpleasant one, and Brown was in the mood to listen to suggestions involving Detroit Edison.

Newspapers at the time suggested that this appointment was dictated by the discontented minority interest, specifically by the United group. If this was true, Brown himself was not informed of it, and in any case it made no real difference. He had no knowledge or technical competence in the field of electric utility as such, but this important part of the Company's operations was in the expert hands of James Parker and his staff. (Parker was elected president of the Company at the time Brown was named chairman and chief executive.)

Brown could and did, however, bring to the Company his great prestige and his knowledge of government procedures and personnel, and he could legitimately use both in the interests of the Company. A decade after the event, he could say, reflectively,[2]

> The number and magnitude of the decisions of government which affect your Company are much greater today than formerly. The emphasis in government during the past twenty years has been on economic and social problems. As the new condition must be lived with, consciously or unconsciously, your board must have had this in mind when they hired a man fresh out of government.

The interrelated problems of finance and the shareholder's interest were specifically in the responsibility of the new board chairman. Brown's legal experience was extensive and, with his

Prentiss Brown.

participation in public and political affairs, he brought a wide knowledge to the service of the Company. This competence he would need, for the interests of the United group as shareholders were being emphatically presented.

The Company had its own financial problems which policies of the past decade accentuated. From 1929 through 1933, as the demands for power slowly declined, generating equipment which was old and less efficient was allowed to stand idle, and some of it was actually dismantled. Even when the city and territory began that long and difficult economic recovery, the demands for power increased only slowly, and the Company output was far below its capacity. In 1935, both the capacity and the peak load were still below the 1929 level.

Without pressure, then, the Company could carry out a steady program of improvement and modernization; the purpose was to attain that measure of economy which new machines offered. Beginning with 1936 and extending to the war's interruption, the Company almost annually installed first 60,000 kw and then 75,000 kw of new capacity. In the depression days, the contract prices for such machines were attractively modest.

The cost of this operation was met with internally generated funds. For ten years after 1935, the Company found impossible anything like normal financial operations—it was fortunate that the financial demands were so modest. The two holding companies, operating under the Holding Company Act, were unable to participate in financing and unwilling to see their interest diluted by the issue of bonds or stock to others. The Company was able to negotiate a ten million dollar bank loan, to be repaid out of current revenue at the rate of ten percent each year. This did not affect equities, but on the other hand it was at best an improvisation.

Postwar decisions made it clear that such fiscal methods had become inadequate. In 1947, the Company planned to spend more money than it had in any year in its history, and it scheduled for the immediate future a capital increase equal to half the total investment. It became essential that the Company aid the United group in the required sales of its shares, in order that the Company could be freed from the restraints which, under the law, were the product of the relationship. Until this happened, normal financial requirements would be met only with difficulty.

In 1947, the board voted to issue a ten percent stock dividend.

This was a belated distribution of a portion of the retained earnings which, under the policy of the fixed dividend, had been accumulating for years.

The stock dividend, helpful though it was, did not affect fully the balance between debt and equity. The stocks represented 59.3 percent of the capital, a proportion higher than normal or advisable. In 1947, a $60,000,000 bond issue (Series J) was sold, in part to refinance Series F to the extent of $30,000,000 but also to finance new construction. Finally, in 1948, a large number of debentures were offered to the stockholders at prices which proved financially advantageous. These debentures were notes of the Company but were later convertible, under specified terms, into Company stock.

Prices of the Company stock responded favorably to such aggressive leadership. In 1946 and 1947, prices had returned to the prewar level, though they were still far below the highs of 1929 and 1930. The holdings of United were quietly liquidated, and in 1948 and 1949, respectively the two men who represented that interest retired from the board.

The stock dividend and the distribution of the valuable rights to debentures as a privilege forecast the first change in dividend rates since the dividend was reestablished in 1936.[3] An increase occurred in 1951, a mere five cents added in the final quarter to the traditional annual $1.20; it was, however, the first of fourteen increases in the next sixteen years. The tradition of the fixed dividend had been abandoned most convincingly.

Detroit Edison had always been a popular investment in its service area, and a great deal of the stock was in the hands of small investors there. In the fall of 1952, the "area stockholders" meeting was called for Detroit's Masonic Auditorium. More than eight hundred people were there, with eight of eleven directors, and the results of this informational meeting were so satisfactory that it was made an annual event.[4] Brown presided at this first one, and in his remarks took note of the Company's financial achievements. Including cash dividends and the valued rights, the Company had distributed above seven percent on its invested capital.

In a thank you note to employes and the Company prior to his retirement, Brown was more specific. In his term of office two blocks of common, two of convertible debentures, six of mortgage bonds and three promissory notes had been issued—and the last bond issue was at less than three percent. The improvement in the

Company's position, plus, of course, the general financial situation, allowed the refinancing in 1954 of 3⅞ percent bonds at a new rate of 2⅞ percent—ultimately saving the Company ten million dollars.

Let Brown himself sum up the achievements of the Company during the decade in which he was its chief executive. "There are few industries that can boast of a record—which roughly, amounts to doubling of the wage level, a reduction of the domestic unit cost of the product, a one-third increase in our dividends and a 60 percent increase in the price paid for our principal raw material, coal, all of which has been accomplished with a background of a 54 percent increase in the cost of living."[5]

Brown liked to tell of his introduction to the Company. He had hardly reached his office when Parker rushed him off to Lansing, to an appearance in a rate case of which he had only just heard and on which he was completely uninformed. The whole episode developed something of the quality of a teapot tempest but, for the moment, the matter seemed anything but humorous to Parker.

The issue was quite simple. During the war, when plants ran three shifts and homes and services often functioned at irregular hours, the amount of electricity used greatly increased. A given machine would be run for more hours a day, at efficient capacity, with only a small increase in cost and with no increase in capital invested. With Company employment down, with construction almost prohibited and with many normal expenditures impossible, the increased income gave an abnormally large profit figure when compared to prewar years.[6] The company found itself guilty of "excess profits" as defined by Federal tax law.

When it appeared that a large sum of money was to be paid to the Federal government in tax on this excess, the local community made an effort to prevent this. The press and city hall combined to urge a "voluntary" bonus to customers either by a return of money or by a period of free service. Either method seemed inexact and probably inequitable, and the Company refused. It was immediately denounced for denying a dollar rebate to the customer because thirteen cents of that dollar would come from the Company itself.

The city took an appeal to the commission in Lansing, demanding an order reducing the rates and, in the meantime, it enacted a hastily drawn excise tax which was intended to secure the funds in question for postwar city improvements.

To add to the confusion, the commission ordered a sweeping reduction in rates and at the same time ordered a large cash refund. Like the excess profits tax itself, the commission's ruling seemed to focus only on the immediate wartime situation, and ignored the effect of its actions on the Company when normal conditions again prevailed.

For the second time in its long history, the Company appealed a rate decision to the court. In the meantime, it segregated the sum in question ($10,450,000) from other funds and from that moment became almost a bystander in the ensuing squabble. The Federal government claimed the money for tax, the city government had its excise claims, and the commission ordered the money returned to the customers in the form of a rebate. The only thing certain was that the Company had no active interest in this gain from its "excess profits."

The court decision confirmed a refund to the customer of the fund which, by December, 1945, had grown to sixteen million dollars. A rate reduction also went into effect; as predicted, when normal peace time demand returned, this cut proved to be excessive. The Company asked for and received a rate increase in March, 1948, which was then estimated at about 11½ percent. It was the second substantial rate increase in the Company's history, and was in considerable degree made necessary by the errors in the earlier ruling of August, 1944.

It was not in the tradition of the company to challenge its customers or fight decisions of properly constituted regulatory agencies, and Parker was apprehensive of the consequences. In fact, no unfortunate result ensued. The issues were clear and the Company was totally candid in its handling of the affair. Over even a short time, there was no measurable loss of public confidence, and that was the major worry.

The involvement of the Company with the Federal government had been slight throughout most of its history. Following the regulatory legislation of the thirties, however, the impact of events in Washington were increasingly felt in Detroit and in the Company itself. Prentiss Brown's leadership in this period was especially fortunate. The achievement of the Canadian interconnection is one of several events in which his unique talents and experience were felt.

The problem of the Canadian interchange is in fact two problems; the one technical and economic, and the other legal and administrative. The technical problem was not complicated. The river boundary between Ontario and Michigan is a little more than a half mile in width. A tower system swinging lines above the water would have to be only high enough to clear steamship traffic on the rivers. The Edison system already operated its tower system at 120,000 volts, which would be satisfactory for such a line.

The Ontario Hydro-Electric Power Commission seems to have made the first suggestion of an interconnection late in the thirties. The Ontario system was suffering the difficulties often associated with one dependent on water power. It was not flexible and could not adjust to peak loads easily; it suffered seasonal variation in water flow, with serious results when high electrical demand coincided with low water supply.

Hydro-Electric was especially concerned with that part of Ontario closest to Michigan, where both industry and population were growing rapidly. This area was farthest from water power, and in any case, the available power was inadequate. The combination of hydro-electric generation with thermal generation would have been the preferred solution but the construction of adequate thermal plants seemed to be out of the question at the moment.

An interconnection with Detroit Edison would provide mutual advantages. When the Canadian plants were producing power beyond their need as, for example, late at night, that energy could flow to Edison at prices set well below Edison costs of coal alone. On the other hand, Edison's thermal plants could help solve the problem of peaking for Hydro, and would greatly improve reliability, especially in the Windsor area.

The administrative problem on the Canadian side was simple. Hydro was so much an arm of the provincial government that its decisions could be implemented easily. As the conversations continued, Hydro made other moves to facilitate interconnection. Though there were other reasons than interconnection for doing so, Ontario Hydro altered its whole system from twenty-five to sixty cycles, which made it compatible with Edison's system.

On the American side, the administrative problem was much

The Interconnection from the grounds of the Delray plant to the lines of the Hydro-Electric Commission of Ontario.

more complicated. Since it was totally intrastate in its operations, The Detroit Edison Company was subject to the regulation of the Michigan Public Service Commission, but not of the Federal Power Commission. There had been a great deal of feeling generated in the Company over the activities of some federal agencies and regulating bodies, and the resolve to avoid such jurisdiction was deeply rooted.

Only after some years did Edison move to examine the request of the Hydro people. The major problem remained how to achieve the Canadian interconnection without the risk of involving federal jurisdiction. The law establishing the Federal Power Commission gave it extensive powers, when commerce was interstate, and the commission itself had issued a rule which defined international commerce in electrical power as legalizing the exercise of federal authority. When the Company's attorneys insisted that the law did not, in fact, authorize such a rule, the commission refused to reconsider or modify its position, and with some arrogance invited the Company to begin a long and expensive law suit if they chose to test the commission's interpretation of the extent of its own power.

The Company, of course, wanted to avoid the delay and expense of such judicial proceedings but with the passing of time the economic urgency increased. In the postwar decade, the mounting Detroit demand and the almost critical developments in Ontario again emphasized the mutual advantage, and the Company determined to move. Walker Cisler was back with the Company after war service which had involved him in just such technical problems. Prentiss Brown brought his knowledge of the ways of Washington where the action would occur.

The course of action chosen was not by way of the courts and its delays, but directly to Congress with the request for clarifying legislation. The argument substantially was that Congress had not given control of international trade in electricity to the commission. The legal campaign was planned by Harvey Fischer, counsel for the Company in Michigan, and by Paul McQuillen representing Sullivan & Cromwell, New York counsel for the Company.

The argument of the commission was that the control of the sales and import of power over international boundary was only an extension of the interstate power in its hands, justified by its reading of the law's provisions. Somewhat more aggressively, the representative of the commission argued that, in any case, it should have the power, and that the use of this device as a means of forcing Edison into Federal jurisdiction was justified.

The Michigan delegation in Congress, both in the House and Senate, gave the Company full cooperation. The Senate bill was presented by Senator Potter, who thus became a participant in the committee hearings. Prentiss Brown presented the chief statement for the Company, giving an historical summary and outlining the issues. A number of others followed, including Cisler who talked of the technical advantage, McQuillen who submitted a brief on the precedents, and Commissioner McCarthy who talked on the adequacy of his Michigan Commission in protecting the public interest.

The Commission voluntarily rescinded the rule establishing its jurisdiction, but this could have been restored by subsequent administrations; therefore, the Company pressed for definitive legislation. Before he left office as chairman, Brown had the satisfaction of seeing this accomplished.

In the summer of 1953, when it was clear that the requested legislation would pass, the preliminary work for the steel towers began. The lines could not be strung until the legal technicalities were completed, but the cables were ordered and ready. Suddenly, an emergency in Canada led the head of Hydro to appeal to Edison for immediate help. Edison explained the situation to the Federal Power Commission by telegram, requested all possible expediting of the final formalities, and received by return wire a full authorization to proceed.[7] In a matter of days, the line was in place, and the circuts closed.

Prentiss Brown brought to the Company the prestige of the public figure. The role came naturally to him in Detroit, for he was accustomed to the obligation imposed on such a public man. For a decade, he was an important leader and participant in the civic and cultural activities of the city and the state. He did these things as an individual, but he was very conscious of the fact that the role affected the position of the Company also.

Alex Dow insisted, as has been said so often, that the interests of the city and the Company were one. Nevertheless, he kept the city and its involvements at arms length. He did give engineering assistance to the Water Board, and his son Douglas aided the Murphy Administration briefly in purchasing, but these were rare exceptions. The autonomy of the Company was expressed in the refusal to join associations, reform organizations, pressure groups or "causes." Dow would not join the employers association or the Board of Commerce, lest he seem to justify the worker joining a union. In Dow's mind, loyalty to the Company was a demonstration of civic virtue.

Both Brown and Parker found this Dow position impossible, and each man in his own way moved to take civic responsibility, and to associate Detroit Edison with their action. For reasons which related to both training and temperament, Parker's activities were chiefly professional in their character.

There was no such limitation to Brown's interest. He maintained his association with his political party, and at least once attended the Democratic National Convention. He was active in his church, and he served on the board of his church's college. His interest in history led to his association with the Michigan and the Detroit Historical Societies, both of which he served as director and president. This historical interest he further extended to membership in the Newcomen Society of North America, an organization which printed an historical address of his on Alex Dow. He was active in and an officer of the Economic Club of Detroit—in his own phrase, he made that club bipartisan.

A list of this sort can only be representative, suggesting the sweep of activity. By the nature of his position in the Company, its routine direction was in other hands. This made Prentiss Brown available. No one will count the committees he chaired, the drives and fund-raising activities he shared, the meetings and causes which gained in importance because he supported them.

This role of public citizen and community leader was especially significant in the postwar era. The civic leadership which had been so important and so generous in Detroit in the decade before 1929 came from that generation which was ruined by the depression. A sense of community entity and obligation had to relearned and Prentiss Brown and a dozen men with whom he worked symbolized that community spirit in Detroit indeed. In 1952, a citation by the University of Michigan said that his "attainments . . . place him in the small company of first citizens of the State of Michigan."

In June, 1950, the newly named Mackinac Straits Bridge Authority held its first meeting at Mackinac Island. To no one's surprise, Governor Williams named Prentiss Brown Chairman of the Authority. The bridge over the Straits, which had been a lifetime interest of Brown's, would now become a reality.

For several years, Brown continued to serve Edison but at the end of June of 1954, almost exactly a decade after his appointment, he retired. He remained a member of the board of directors and was retained as consultant on legal matters.[8] His retirement was the signal for overwhelming expressions of gratitude and affection, not only from within the company, but from the city itself and the people he had served so generously.

For Brown, retirement simply meant that he could return to the Upper Peninsula and devote more time to the "Big Mac." In 1958, when the bridge was dedicated, the board of Detroit Edison met at Mackinac City for its monthly meeting. Thus, officially it took recognition of an achievement, so much the work of one of its members.[9]

Chapter Four

THE COMPANY AND ITS PRESIDENTS 1944 - 1954

James W. Parker and the Company / The Memory of the Depression / Parker and the Postwar Construction Program / The Parker-Cisler Decade / The Postwar Worker / Retirement and Summary.

When A.C. Marshall retired in 1944, the board divided the functions he, and Dow before him, had performed. The position of president of the Company, as distinguished from the chairmanship of the board, was given to James Parker, who at the time of his appointment, was vice president and general manager. There is no question but that many men inside the Company would have been happy to see Parker made Marshall's direct successor, for his competence was recognized and admired. As vice president under Marshall, he had had responsibilities beyond those in technology, but he was not well known outside the Company and the profession.

Brown as chairman of the board was the chief executive of the Company, with his direct involvement chiefly in the areas of governmental and civic affairs and in financial and legal matters. Parker as president was responsible for engineering and for technical planning and administration. Between Brown and Parker a confidence resting on mutual respect grew rapidly.

James W. Parker had had a distinguished career in engineering. He had been with the Company for thirty-three years and during that time had accepted constantly mounting responsibilities. More than any man since Dow, he had grown with the Company.[1]

Parker came to Detroit Edison directly from the Engineering College of Cornell University in 1910. He was not the first college trained engineer hired by the Company, but he was the first to take his technical training directly into the power house. In the rough and often crude fraternity which by tradition ruled there, Parker was of a new breed, but he quickly won respect and acceptance.

He entered the Company at that moment of transition, when the trial-and-error, rule-of-thumb methods of the old school plant superintendent were giving place to the use of scientific equipment and calculated measurements. One disgusted veteran called it "power house bookkeeping." As a young engineer, Parker shared in the application of scientific techniques and he recorded his important findings in the journals and reports of his profession. But Parker knew, too, the importance of men as leaders, and at his subsequent insistence, the road of advancement was kept open to those competent men whose education was that of the apprentice and who were graduates of the power house. It became a Company tradition that the new college trained engineer serve his time in the plant.

Parker's career in the Company appears to be erratic only until the pattern emerges. Dow and Marshall used him where there was trouble. He became vice president in 1935, just as labor problems became increasingly critical. More and more they became his special concern. During the prewar preparedness program and the war days themselves, he had worked closely with Marshall. Now as president, he must meet the demands of the continuing war emergency, and anticipate the postwar requirements of the great area which Edison served. In technical training, in local experience and in familiarity with the city and job, Parker was ready for the new responsibility.

Central to the Company were the four power plants located at intervals along the fifty mile water way between Lake Huron and Lake Erie. At the opening of the postwar period, the machines in these four plants had a generating capacity of 1,233,800 kilowatts. In January, 1944, the wartime peak demand had been 1,055,000 kilowatts; the margin was a safe one and seemed to provide assurance for the immediate future.

The territory in southeastern Michigan which these four plants served was roughly the size of New Jersey or Massachusetts, and had a population of a little less than three million people. The balance of Michigan had a population also of roughly three million, and was served by two major companies and a number of local, municipal or cooperative organizations.

To serve so extended an area, the Company had developed before the war a distribution system of strength, economy, and reliability. A loop stretched from the Marysville Plant on the northeast to Trenton Channel on the south. Lines between individual plants on the river side included the other two older generating plants, Delray and Conners Creek. The reserve of each of the plants was thus available to support the whole system, and the entire Company could be operated centrally with economy and security.

The idea of mobilizing existing surplus of one plant as a reserve available to another had obvious possibilities for extension beyond the lines of the Company itself. Such an interconnection was established with several smaller industrial generating plants, including both Ford's Rouge Plant and the University of Michigan's campus plant, but also more importantly with the lines of the Detroit Public Lighting Commission and with Consumers Power. Both in the power house and in the field, Detroit Edison equipment was maintained meticulously, with a long standing program of depreciation and replacement which kept it up to date and efficient.

To both the men in Edison and the leaders of the City of Detroit in 1944, the economic disasters of the depression years were recent history. The depression hit Detroit earlier than it did the rest of the nation, augmented by the prolonged shut down of Ford's plant while engineers designed and produced the Model A car. From 1927, despite the external appearances of prosperity, unemployment was a mounting fact in Detroit's labor community. When the rest of the country began to suffer depression, Detroit's disaster became complete.

The first five years of the next decade found everyone floundering in massive and seemingly endless frustrations and defeats. In

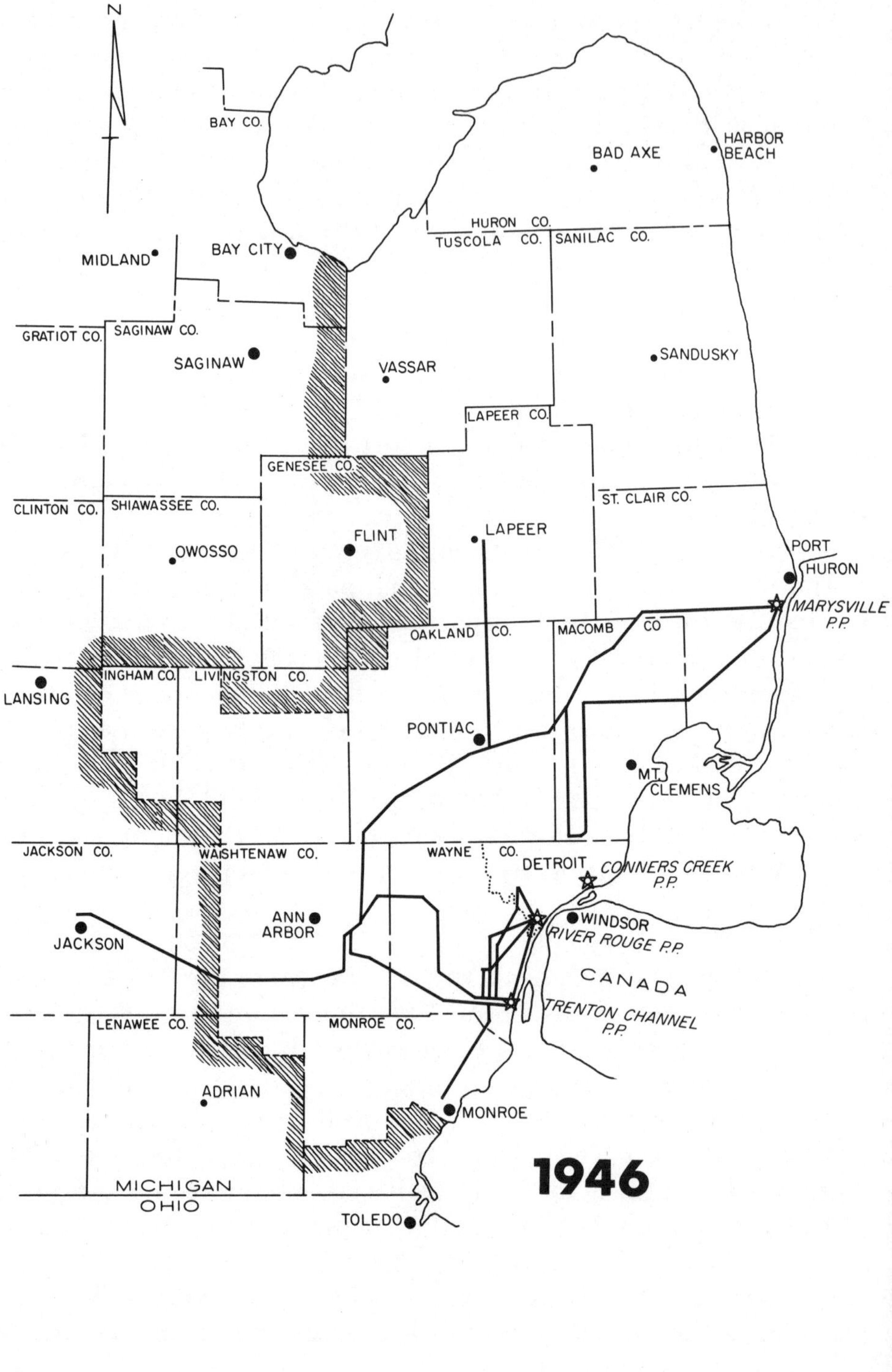
N
BAY CO.
HARBOR BEACH
BAD AXE
HURON CO.
TUSCOLA CO.
SANILAC CO.
MIDLAND
BAY CITY
GRATIOT CO.
SAGINAW CO.
SAGINAW
VASSAR
SANDUSKY
LAPEER CO.
GENESEE CO.
CLINTON CO.
SHIAWASSEE CO.
ST. CLAIR CO.
FLINT
LAPEER
OWOSSO
PORT HURON
MARYSVILLE P.P.
OAKLAND CO.
MACOMB CO
INGHAM CO.
LIVINGSTON CO.
LANSING
PONTIAC
MT. CLEMENS
JACKSON CO.
WASHTENAW CO.
WAYNE CO.
DETROIT
CONNERS CREEK P.P.
ANN ARBOR
WINDSOR
RIVER ROUGE P.P.
JACKSON
CANADA
TRENTON CHANNEL P.P.
LENAWEE CO.
MONROE CO.
ADRIAN
MONROE
1946
MICHIGAN
OHIO
TOLEDO

the Company, fewer domestic customers were served in 1934 than in 1929, fewer persons were in the area (by Company standards) than in 1929, and the revenue of 1929 was not equaled until 1937.

No one really knew exactly how many were unemployed but the guess was that one-fourth were totally without work, and a considerable part of the remainder were on shortened hours, share-the-work programs, and cut wages. Bread lines were pathetically long, welfare hopelessly inadequate, and relief efforts had bankrupted private charity and local governments as well.

The Federal government's attempts to reestablish the economy as a going concern through the use of artificially contrived purchasing power were not well received. "Pump priming" was a term suggesting the theory involved. In 1936 and even more in 1937, the rising level of activity gave promise that the economic system was again self sustaining. The Federal government hopefully cut the priming, as theory had proposed, and immediately the economy went into a decline more abrupt than the one in 1929. To many observers in postwar Detroit the war preparation and the war had been, in the economic sense, giant WPA programs, the effect of which might well be temporary also.

The recollection of these facts and fears affected men as they looked toward the postwar period. Prewar and wartime demand for laborers brought perhaps as many as 400,000 persons to Detroit by 1945; the population rose from less than two and a half million (1936) to over three million in 1944 in the Company area. Obviously these people, insofar as they were adult workers, were chiefly employed in war-related industry on contracts which would terminate abruptly. If Detroit returned to its prewar level of civilian employment then the future for these persons recently come to the area, and for the men returning from military service, was indeed bleak. One local organization estimated that there would be nine men competing for five jobs. In this atmosphere of uncertainty, someone had to plan for the future of The Detroit Edison Company.

Parker brought to the presidency in 1944 a detailed and intimate knowledge of the Company in its technical and physical aspects. This had been his job as chief engineer since 1925. In this position also he had to deal with problems growing out of the boom and bust economy, especially as it affected production and construction. As president, additional responsibility for the decisions concerning the postwar years would be his.

The arguments of the pessimist he knew, understood, and from the first frankly rejected. Parker led a campaign in the local Board of Commerce and among his fellow business leaders, advancing his views. He and his committee sent a massive questionnaire to employers of the area asking not for predictions on general economics but for a statement of personal plans, and discovered that most of the men expected or planned that their industries would convert to civilian uses rather rapidly, would hold their employes, and that many expected business to grow materially.[2] Parker himself, looking at the Edison Company, said that it would accept all its employes in service who chose to return and he estimated that additional employes to the number of four hundred would be needed. Remembering the experience after World War I, Parker calculated that the peak demand would come in the postwar year or years.

This forecast, if it was to be meaningful, had to be backed up with action and investment. Parker, with Paul Thompson, worked out a plan which Prentiss Brown and the board approved. They announced a ten-year program of construction which would cost $125 million. It was a bold announcement, by a Company whose total capital at the time was less than $280 million. It was also bold considering that even war demands had not strained the Company's existing capability.

Plans were made to install immediately at Marysville the 75,000 kw machine which had been war-delayed.[3] The machine came on the line early in 1947 and, with that installation, the capacity of the Company was elevated to 1,309,000 kw. It was just in time, for the surging postwar demand in 1947 lifted both total production and peak load well above the war maximum and 1948 was sharply higher.

Hastily, the Company developed plans for substantial additions to two plants. During the depression, as a part of the modernization program, Conners Creek and Delray No. 3 had been substantially rebuilt. Additions were now planned to the Trenton Channel and Conners Creek Plants.

The new addition to Trenton Channel, begun in July 1947, was substantially a new plant, housed in a building roughly the size and shape of a football field. The two generators, No. 7 and No. 8,

The older Marysville Power Plant, shown with the addition completed soon after World War II.

The Trenton Channel Power Plant, showing the additions of 1949-50.

each 100,000 kw, increased the capacity of the plant to 500,000 kw. By 1949, this plant was close enough to completion to allow attention to go to the extension at Conners Creek. There, two substantially identical machines were installed, rated at 135,000 kw each.

But construction still did not satisfy the rising demand of postwar Detroit. As December approached, with its inexorable peak demand, the Company worked frantically to get every machine prepared to produce the maximum kilowatts. Year after year, despite all efforts and even with new installations, the peak strained the Company beyond its theoretical or name plate capacity, to test that built-in safety margin each machine carried. Yet almost annually, the president was able to issue a congratulatory statement to the men of the Company, and at least once he boasted of their common achievement in a notice in the local press.[4]

On August 1, 1951, turbo-generator No. 16 was turned on in Conners Creek. This was the second of the two machines scheduled for this structure and it marked the completion of the rebuilding process carried forward since the war.

Parker must have been a little amused when he recalled the incredulous reception given his estimate for postwar Edison. He had estimated that an investment of $125 million would be needed in ten years. In fact, before he retired in 1951, the total amount had been spent in five years, and in the ten years the expenditure was twice as much.

Even Jim Parker would not have guessed that the peak load of 1940, inflated though it was by preparedness activity, would be nearly doubled in a decade, or that the census of 1950 would show the Company area with a population increase of twenty-five percent. The nation's increase had been only about fifteen percent. By the end of 1952, the increase in customers (230,000 since the war) meant the growth in population equivalent to a city of 700,000. There were only ten cities in the nation whose total population exceeded this Detroit area increase.

Parker remained in the office of president from his election in 1944 to 1951. In 1948, Walker Cisler was named executive vice president. In this post he had to deal with power plants as they struggled to achieve adequate production, and with labor as it sought status and recognition. In all matters he worked closely with Parker. When, in 1951, Parker retired from the Company, the appointment of Cisler as his successor made the transition in office a smooth one. Cisler, rather naturally, tended to maintain the policies for which in major part he was himself responsible, so that the Parker-Cisler presidency may be considered as a continuing administration, at least until Brown also retired.

The battle to meet the rising demand for energy was probably a first concern with the management but second only in degree of urgency was the problem of employe relations. This was a totally new situation in collective bargaining, for organized unions now had enforceable rights under Federal labor law. This problem, too, fell into Jim Parker's hands, first as vice president under Marshall, and then as president of the Company.

Parker's long career with the Company might well have been designed to prepare him in both experience and personality for what was substantially a crisis in labor relations. For a short time he had been in charge of the heating division of the Company, a small and in many ways, autonomous unit. Here a sympathetic man could get to know his people and understand not only their problems but their prejudices as well.

For a brief time, when the Company attempted a not too fortunate experiment in owning and operating captive coal mines, Parker was directly in charge. The experience with the labor and general living conditions in the southern coal areas was one which he could never dismiss. He was president when he said that he did not oppose unions so much as he opposed the conditions which made unions necessary.

In many ways, Parker belonged to the older order in the Company; he had worked a long time with Dow and they were exceedingly close both personally and professionally. He, too, would no doubt have preferred the simple paternalistic labor relations, with its system of unilateral decision. He knew, however, that this system was gone—if for no other reason than that there were no longer twenty applicants for every job available at Edison.

Unionization brought difficulties to both sides of the labor controversy, and the line of division in the Company was not the classic employer-employe one. Some employes were more uncompromisingly opposed to unions than were their bosses; they resented their fellow workers who purported to speak for all employes, and they joined management in expressing distaste at outside interference.

Unfortunately, much of this loyalty had been personalized in the old leadership. Dow was still alive when gossip began that "the Company isn't what it used to be." He came out of his not too complete retirement to defend Marshall and Parker, but the talk continued. It would be unfair to Marshall, and especially to Parker, to view the Company employe program of the war and the postwar period as a mere popularity contest with the union organizers. The long standing paternalism had reflected an honest concern with the well being of the worker; the discovery that any sizable body of the workers was discontented was a surprise.

Parker did not like unions. He shared the Company prejudices about organizers and their motives and was deeply hurt when the election of the employes first showed that they preferred "the leadership of such men" to that of the Company and its executives.

He did not like unions but he accepted them, and imposed two injunctions upon his subordinates, including supervisors at the lowest level. (1) We do not fight unions. (2) We insist that those in unions and those out of unions are both employes, and should receive the same consideration. The responsibilities of the Company were not limited to the terms of the contract itself, even to union employes.

Employe relations, then, were vastly broader than the dealing with unions and contracts, but the unions were new. Under the law, a petitioning body of workers had the right to an election, the issues of which included the decision on the matters of unionization itself, and the choice of the bargaining agency. The law required the recognition by the Company of a bargaining agency so established, and defined collective bargaining.

Division inside the Company among employes made the majorities either way small and the results unstable. During the war itself, when wages were in theory frozen, the threat of a strike could sometimes attract the attention of the War Manpower Board and win approval for a wage increase otherwise impossible. This proved to be a strong argument for the organizers. By the war's end, the total membership under union contract was between one-third and one-half of the operating employes.

In the decade which followed that first union contract (1943) conditions did not change much. The major unions were the International Brotherhood of Electric Workers #17 and the much larger and less homogeneous affiliate of the CIO, the Utility Workers Union of America #223. Some divisions of the Company were totally organized or near it, while some divisions were totally non-union. By 1955 roughly half of the workers were working under contract.

The threat of strike is, of course, the ultimate argument in negotiation. In 1952, Local #223 called a strike of its thirty-seven hundred members, and was out for twenty-three days. In 1955, Local #17 struck and was out for thirty-four days. Substantially the issues involved pay scales and, though there were some differences, both contracts were written to incorporate the principle of the cost of living adjustment.[5]

Whether negotiated in strike or, as was much more usual, in less dramatic confrontation, the terms and benefits of the agreement were automatically and immediately extended to workers not covered by the contracts. Perhaps unconsciously before, but specifically after 1950, the wages paid by the Company were directly related to those paid in the general Detroit area.

Very soon after the war the Company became concerned with the question of employe attitudes, and Parker secured the services of outside consultants to conduct an exhaustive survey. They found some curious contradictions; for example, employes liked the job security, but they resented the slowness of promotions. The Company could do little about that dilemma.

One cause of concern was the employe attitude toward the retirement system. This had been approved in 1943, as a formal plan to take the place of the "grants" which sounded so much like charity. The pensions were totally at Company expense and were in addition to Social Security, but there were confusing references to "prior service," limitations to the age of thirty and after, and other misunderstandings on terms and value.

Walker Cisler as executive vice president undertook an extensive series of departmental and divisional meetings, and brought back reports which led to a reexamination of the contract. On its own part, the management was concerned with the sharply rising age of employes, and with the effect it would have on the solvency of the fund itself. In 1950, more than two thousand employes, almost one-fourth of the total number, had been with the Company for twenty-five years.[6]

In 1950, Parker announced his intention to present to the board a proposal for a modification of the pension plan. It was a part of a series of changes which looked toward both improvement of the benefits and clarification and simplification of the plan. Total time worked, not just that after the age of thirty, was to be credited and thus the old maximum of 35 percent of pay was removed. By 1956, the average employe was retiring on 52.4 percent of final pay, including Social Security. In that year, the trust fund totaled $49 million and the number of retirees on the role was up to 1,427.[7] The cost of the pension fund was sharply increased. Cisler in his first year as chief executive officer (1955) stated that the total fringe benefits plan cost the Company $12.7 million annually, or about 56 cents per hour worked by employes.

The survey of employe opinion had disclosed also a good deal of dissatisfaction with pay schedules, especially as they differed between departments. There was complaint that men competent for promotion were passed over for irrelevant reasons. The survey seemed to challenge the whole system of pay and promotion. Parker proposed and initiated a system of job evaluation which was Company wide and which attempted equity.[8]

The complaints often were vague and obviously personal but they justified a reexamination of basic employment and promotion policies, and a more rigorous and formal search for competence when job opportunities opened. In 1955, three years after the establishment of the system of posting vacancies and promotion opportunities, the older complaint could be answered with the statement that among employes that year, one in six had been promoted or transferred in grade by the system of job bidding and that, during the three years it had been in effect, there had actually been 409 interdepartmental shifts.[9]

War and the long extension of the draft created a continuing problem for it interfered with the normal processes of recruitment and job training. When men in the Company were called to guard duty or to reserve summer training camps, the Company did not charge that time against vacation and it supplemented the service pay to make it equal to regular compensation. To the extent that it was possible, men called to full war service retained seniority against the time of their return.

One wartime improvisation became Company policy. A rule, so old that it predated the Company, forbade the employment of married women. When a girl married, she severed her connection

The Conners Creek Plant, a long time landmark, was rebuilt during the thirties. The extension shown is from 1951.

with the Company, although a "marriage grant" of one week's pay for each year to a maximum of six was given to the bride.

The war increased the number of women employes and the significance of the work some of them were doing. The rule seemed outmoded as well as inconvenient, especially with so many husbands away on duty. It was suspended for the war, reinstated with considerable difficulty against the protests of both the women and their bosses, and then questioned again in the Korean crisis.

In 1952, both the rule relating to working wives and the provision for the marriage grant were permanently suspended. Since the women, if they so chose, could get jobs elsewhere for the asking, continuation of the rule would merely cost the Company valued employes with skill and experience. Oddly and unexpectedly, the protest against the abolition of the rule came from other women. Dismissal of married women had resulted in rapid promotion for the single girls. They would rather, it seemed, tolerate the "discrimination" and enjoy the advantages in pay and position.

During the war, morale had been deliberately cultivated. The efforts on the part of the management afterwards to maintain it at a high level and their methods would perhaps surprise an outsider, but the purpose was to produce a happy and contented employe, and that was good. Employe clubs were of long standing and were only slightly subsidized, but they held together hobby enthusiasts in bridge, photography, radio, chess and checkers, and fostered a dozen other activities. There was an Edison American Legion, with a prize winning Bugle and Drum Corps. Athletic events and contests included golf, soft ball and sometimes interplant series, or even participation in an industrial league.

The old house organ, *Synchroscope,* which had been so sedately instructional in content, was now devoted directly to employe interests and activities. With the expansion in size and complexity of the Company, the development of the picture reporting techniques and articles on Company plans and developments gave the magazine a new significance. To further stimulate pride and interest in their work at least one television show was directed to employes, a movie called "My Dad's Company" was produced, and "open houses" for employes and their families were held on all possible occasions.

If some of these activities seem slightly synthetic, it is because some of them were. In the main, however, the Company behaved with self-respect, treated its employes with honesty and made every effort at fairness. It did not substitute activities or propaganda for solid recognition.

Occasionally men in management felt they were subjected to irresponsible complaints. In 1954, Syl Leahy, whom Parker had made Director of Employe Relations, responded to an employe who questioned the "value" of the whole program with an impressive list of employe benefits.[10]

1. Retirement pay totally at Company expense, Group Life Insurance partly subsidized and pay for normal absences for illness.
2. Vacation; 6,724 employes for two weeks, and 4,058 with 15 years seniority for three weeks. Holidays with pay, seven and one-fourth days.
3. Hospitalization and surgical insurance for 30,000 employes and dependents. Last year 6,300 persons benefited.
4. Layoff allowance after six months employment.
5. Employe clubs for professional and recreational purpose. 8,000 members in one or more of these partially subsidized clubs.
6. Social Security with partial unemployment compensation. Medical care and pay in the event of occupational accident or illness.

Leahy refused to attempt to put a cash value on the total, but he underlined the point he wanted to make. All in all, the Company was a good place to work.

The steadily rising average age, which continued from the decade of the twenties to the fifties, was an indication of continuity of employment. One out of four new employes had relatives already working for the Company. A surprisingly large proportion of men who left the Company for duty either in World War II or in the Korean conflict returned to Company employment.

In the last five years of Parker's presidency, the number of employes owning stock doubled; the total ownership was not large, a mere 3.3 percent of the stock outstanding, but it exceeded the national average of employe ownership in industry and, if this holding can be considered as a block, was the largest concentration of Company stock ownership.

James Parker was among the scientists who recognized the importance of atomic research for other purposes than the manufacture of the bomb, and he was among the businessmen who organized to persuade the government of the desirability of cooperative efforts. In 1947, David Lillienthal, then chairman of the Atomic Energy Commission, named Parker chairman of the Industrial Advisory Group of the Commission. In 1948, this group strongly recommended that industry be given more opportunity to work at the development of energy application—and that industry be required to share in the expense of such development.

Before Parker left the presidency, the Detroit Edison-Dow Chemical agreement was announced (1951) by which the two would launch a program of research to find industrial applications for this new source of power. This was the beginning of Edison's long involvement in atomic power programs. To many people this achievement by private enterprise marked the climax as well as the end of a long career—certainly it was a deep personal satisfaction for Parker.

Parker's professional attainments in the field of engineering had long been recognized. He wrote extensively, but chiefly before he became so totally absorbed in administration; throughout his career he spoke frequently at professional organization meetings. During World War II, he had been on assignment in Washington. After the war he accepted an assignment which took him to Germany, to advise on postwar organization.

In retirement, Parker's energy, and appearance belied his age. He remained active in the Company, for a time as consultant and continuingly as director, and attended a meeting of the board only a few days before his sudden death on December 20, 1957.

The minutes of the board of directors of January 20, 1958, note with expressions of regret the death of James Parker. Special mention was made of the warmth and human understanding of the man and of his part in shaping the relations of the Company with people—employes, customers, and labor unions.

OPPORTUNITY AND THE NEW LEADERSHIP

Introduction

The first half century of Detroit Edison's history ended in 1953, with appropriate notice. For the historian, unfortunately, events seldom conform to such orderly limits, and the boundaries which he sets so neatly can prove to be contrived. Events in progress do not pause to note a date, and the men who served in the Company noted no change in task or prospect.

Circumstances do change, however. The problems which had brought the Company so close to crisis early in the forties had been resolved by the men then chosen to lead, and these events were a part of the history. Perhaps it should be argued that the older regime ended not with the retirement of veteran leaders, but with the solution of the issues with which they had struggled.

It was clear at that time, however, and it became increasingly obvious in historical perspective that the city and the Company had moved to new opportunities, and to the new problems which they created. The operation of a complex industrial organization is always demanding, and sometimes difficulties seem to be compounded. The long depression and the long war which followed created an imbalance. The problems which the Company faced in the two decades which followed World War II were not the old problems, for which solutions were known. The new administration would have ample opportunity to prove its competence.

CHAPTER FIVE

THE NEW ADMINISTRATION 1948 - 1958

The Choice of President / The Posture of the Board / The Emergence of a New Administration / The Council: Duties and Responsibilities / Staff and Recruitment.

THE TRANSITION FROM THE OLDER LEADERSHIP TO THE NEW OCCURRED between 1948 and 1954. It involved the executive officers, the board, and the administrative leadership within the Company. The most dramatic illustration was the designation of Walker Cisler as president in 1951; he was a relative newcomer within the Company, and his election violated the tradition of internal succession in office which had characterized the older regime.

Cisler was not, however, totally new to either the Company or its leadership at the time of his election. As executive vice president since 1948, he had been heavily involved in planning and administration. His relationship was especially close with James W. Parker. Perhaps there was something of the old school tie in this friendship for both men were products of the Engineering College of Cornell, though of different student generations.

Walker Cisler was a native of Ohio but grew up in Pennsylvania with an almost classic "log cabin" background of genteel poverty and the farm. His college work was interrupted briefly by World

War I, in which he won an officer's commission and his career began postwar with his employment by an eastern utility company.

Called to Washington in 1941 by the national emergency he accepted assignments in the somewhat confusing succession of organizations which attempted to direct America's industrial power to maximum war potential. The electrical industry, highly individualistic and independent, found war restrictions irksome, even when it accepted the necessity for them. The limits of the supply of copper, machines, lines and labor demanded an informed rationing both to and by the makers of electrical power and all of Cisler's tact was needed as he dealt with such critical problems. During this time Parker and Paul Thompson from the Company first encountered Cisler and agreed that they would like to bring him to Detroit. A. C. Marshall, wartime president of Edison, persuaded him to accept the appointment.

At almost the same moment, however, the military situation abroad changed and there were critical new needs. Cisler was called into the army—an Edison employe on leave—received his commission as Lieutenant Colonel and was assigned to serve at Allied Headquarters. It was clear that the armies, in spite of the military preferences, must deal with civilians and cities as well as enemy armies. Cisler was the expert whose function was to restore as quickly as possible the electrical service so essential to civilian life and to army efficiency. His assignment in North Africa led him to south Europe with the invading army, through Italy, and he was in France promptly following the Normandy invasion. After the war, he served as advisor to the armies of occupation and to the postwar military governments.

By the time Cisler was ready to return to Detroit, he had earned citations not only from his own government but also from the governments of Britain, France, The Netherlands and Belgium. It is undoubtedly true that at that moment he was better known in Europe than in Detroit. It is also true that his informed and vital interest in Europe and its people did not cease with his release from duty there.

Returning to the Company as an engineer in charge of power plants (1946) he was caught in the drive for production to meet the needs of surging Detroit industry. As executive vice president (January, 1948) he continued to share with Parker the worries over peaks and capabilities. Planning was of necessity short ranged until the manufacturing companies could catch up on the backlog of the war. The Korean crisis extended in limited degree the restrictions of wartime, and revived the problems of war production.

For Cisler's first four years as president, he shared leadership with Prentiss Brown, who continued as chairman of the board. Brown had had a distinguished career in politics and public affairs and Cisler, conscious of the importance of this public position, worked with Brown closely. Cisler had a deep respect and admiration for the older man. In 1954, to anticipate events a little, when Prentiss Brown retired from his post as chairman of the board, Cisler, as president, was named chief executive officer of the Company. The title chairman of the board was allowed to lapse for the time being. This was Cisler's decision and was his gesture of respect to Prentiss Brown.

In the theory of the law and the myth of the business world, the board of directors of any corporation exercises extensive power. The resources of the company are in the hands of the board. The board elects the officers, maintains or replaces them, and accords to these officers as much freedom of action as it chooses.

In theory, the board is elected by the shareholders of the Company, but that is part of the economic myth. In normal practice, the slate of the management and board are presented to the shareholders in the proxy statement, and they are elected by proxy votes. Since the conflict with the United group and the compromise settlement (1944), there had never been a serious effort to challenge the right of the board to name its own succession. It was and is, in fact, a self-perpetuating body. This being true, an analysis of the composition of the board, and an examination of the nature of the changes which occurred, may be informative. Again the period of present concern is the half decade beginning in the year 1949.

The two veterans of the board were from New York. B. A. Tompkins was an officer of the Bankers Trust Company, the financial agency in New York which the Company used; his presence on the board since 1923 was a recognition of this relationship. James F. Fogarty was first chosen to the board in 1920, though earlier his position with The North American Company, of which he was an officer, had given him a close association with Detroit Edison. In a moment of near crisis (see Chapter Two) Tompkins had suggested the election of George L. Burr, a retired officer of Guaranty Trust Company, with extensive business and financial experience in New York.

John Foster Dulles had served on the board while he was the assigned representative of the New York law firm which Edison retained. The complications of the regulatory acts and the rising shadow of international conflict combined to force Dulles' resignation (1940), but after a brief interval a former partner recently retired from Sullivan & Cromwell, Reuben B. Crispell was named to the board. Crispell's election dated from 1943, but the Company had been closely associated with that firm since the days of its New York incorporation. Tompkins, Fogarty, Burr, and Crispell were all from New York, associated with either the legal or the financial concerns of the Company there.[1]

A fifth member of the board, Ralph T. McElvenny, quite frankly represented the interests of the United Group (see Chapter Two) though he was at the time a well-known Detroit businessman. The termination of that interest and his resignation in 1949 signaled the end of that old controversy dating from the holding-company days.

Three members of the board represented the direct Detroit Edison tradition. Russell W. Symes was an experienced accountant who for a decade before his election to the board (1945) had been secretary to the corporation. James W. Parker had joined the board in 1940, several years before his promotion from vice president to president. Prentiss Brown joined the Company and became member and chairman of the board in 1944.

Here, then, were two major "interests" which the board could be assumed to represent. Detroit Edison was a business concern, with obligations to shareholders and employes. It had long and proudly insisted that it was also a local institution with meaning and responsibilities to the community. This tradition of close local association and accountability seemed to be contradicted by the board's membership list.

That Detroit association was in 1949 represented by a single member on the board. Charles T. Fisher, Jr., had been appointed (1947) to replace his father, who had resigned for reasons of health. Charles T. Fisher, Sr. was one of the famous automotive brothers, and both men were also actively interested in the National Bank of Detroit.

Earlier in 1949, the authorized membership of the board was increased from nine to eleven. The resignation of McElvenny and the two new positions created thus allowed the naming of three men. It is not surprising that the three men who were elected were strongly related to the local business community. George R. Fink was the president of Great Lakes Steel, and Robert R. McMath was a local industrialist and scientist. George J. Burke was a well-known attorney and political figure from Ann Arbor. Both in the choice of the men named, and in the decision thus to emphasize the local responsibility, Prentiss Brown's firm leadership can be detected.

As contrasted with the troubled half decade following 1939, the membership on the board was relatively stable. Seven members of the board were veterans with a minimum of five years board experience. In the years 1947–48, four men were added to the board, all from the Greater Detroit business community. In 1950, George J. Burke died, and in 1951, Walker Cisler was elected to the board. With this single exception, the board of 1945, with the additions from the local community, served to December, 1955. This board, as it stood, was a distinguished one. The eleven members, who by design represented the several interests involved, were selected to give strength and assurance.

The relationship between the board and the administration was close and confidential. The formal meetings of the board were scheduled in the fall for the entire year, normally at monthly intervals, though occasionally there would be omissions or postponements. The meetings normally were held in Detroit, in the council room of the Company, but when justified, were transferred to New York or another location. In general, the formal meetings were brief, and the minutes record decisions and actions only.

In 1954, the president introduced a new procedure. Some days before each meeting, an elaborate letter went from Cisler to the members of the board.[2] Routinely it began "The following is a report on current matters which I believe you should have before our meeting . . ." Following this, in six to fifteen pages, were the statistics for the current month, both fiscal and operational, comments on business conditions, especially as major customers were affected, matters of labor relations, government relations, transportation and fuel, and other factual data.

Included also were discussions of matters currently at issue in or outside the Company—especially those matters on which the board might be asked to act. Cisler used the directors' letter also to get material to the directors which he thought would or should interest them. This included reprints, speeches, clippings, programs, magazine articles, and personal correspondence.

This rather formidable communication, with its supplemental material, conveyed the Company's position and concerns, as these were seen by administration, to the directors. Because of its semiofficial and confidential nature, and because of its relevance to the constantly changing business and civic scene, the letter to the board on Company affairs is now an almost unique resource to the historian of the Company. It must have been invaluable to the men to whom it was addressed.

Advancement in the Company, had by no means been exclusively on seniority, but that had been an important consideration. The system of internal promotion and continuity in tenure had sometimes brought up men of limited formal training, but with ability and an intense loyalty to the Company which had given them opportunity and a career. As a matter of policy, promotion was open to all, without regard to formal education, and the Company was the richer for this decision. The problem in 1951, was that too many men were of the same generation. This was true in top administration and also among key people on technical assignment.

The title of executive vice president had been created for Cisler's rather unique role in the Brown-Parker administration. On Parker's retirement, the title Cisler had held became essentially an honorary one, held briefly by several veteran employes as a climax to their careers. In thirty months, three men were successively named executive vice president; for each, this honor anticipated retirement.

James Lobban had had a long and creative career in the Company. He had had a major part in establishing the Company's pioneering property records system, and his subsequent work in the accounting field had helped shape utility practices. Within recent years, he had also been secretary of the Company after

Russell W. Symes' retirement. Paul Thompson was an engineer with an unusual bent toward practical research. With Parker, he had directed and reported on a number of boiler tests of special importance, and he had shared in engineering both machines and the power plants to house them. Arthur S. Albright had had a more varied career in the Company, including a wartime responsibility for research. Prior to his appointment to the executive vice presidency, he had served as controller and treasurer of the Company. These were able men, and the appointment to this high post was well earned.[3]

Eventually the office of executive vice president again became more than an honorary position. When Prentiss Brown retired from his position as chairman of the board and chief executive officer of the Company on July 1, 1954, Cisler became the chief executive without changing his title as president, and Charles Landrigan as executive vice president was thus second in command. His retirement, which terminated a distinguished career, was announced in 1961.

Thomas E. Hurns, an employe since 1909 and a veteran of the accounting department, had been Lobban's assistant, and became his successor as secretary of the Company. After 1955, he was both secretary and vice president. Hurns retired from the Company in 1958.

Cisler attributes one important administrative innovation to his military experience. For many years, the Company had published "organization lists" which indicated which divisions or departments reported to each man and, in general, what responsibilities he carried. In 1952, the title was changed from *Organization List* to *Table of Organization,* and the sub-responsibilities were also clearly designated.[4]

In the history of the Company, the title of general manager which Dow held had indicated his responsibility for the actual running of local operation. When he in turn wanted clearly to identify responsibility as well as succession, Dow gave the title to Marshall. Parker was also general manager, and as president he continued to be in charge of the functioning of plants and systems.

In the 1952 *Table of Organization,* the title of manager reappeared, but with a new significance. As reintroduced, the word referred not to a single person, but to a level of authority. Cisler's expressed intention was to decentralize management; the authority for decision which formerly had resided in one general manager was now to be shared, within established limits, by an especially able group of younger men who bore that title.[5]

In this new listing, Chester F. Ogden was manager of Purchases, Edwin O. George was manager of Sales, Howard P. Seelye was manager of Engineering, William W. Williams, manager of operations, and George A. Porter, manager of Construction. Sylvester F. Leahy was designated director of employe relations, but soon thereafter these responsibilities were divided. Leahy continued in employe relations, Donald Kigar took charge of the union relations and both men were made managers.

The position of manager gave a man the opportunity which might win further promotion as retirement or terminal promotion brought changes in the ranks of the senior officers. On the lower levels also, the establishment of the recognized chain of command in the new *Table of Organization* was followed by a number of retirements, transfers, and promotions.[6]

The total effect of the combined organizational and personnel change was to surround Walker Cisler with an administrative staff of his own choosing. In the five years from December 1, 1952 to December 1, 1957, the Company elected nine men to vice presidency. Hurns and Landrigan were among the nine, but they were veteran employes and soon retired. The remaining seven became the new administrative group. No vice president remained in office who had been elected before 1952.

George W. Porter was named vice president for construction and production in 1952 when Paul Thompson was promoted. Eldred H. Scott succeeded Landrigan as controller in 1953 and became vice president in 1955. Edwin O. George succeeded Fred Compton as vice president in charge of marketing and sales in 1956.

In 1957, four vice presidents were appointed: Arthur S. Griswold who continued as administrative assistant to the president; Donald Kigar, union relations; Sylvester F. Leahy, employe relations; and Chester F. Ogden, purchases and real estate. In that same half decade new appointments were made to the office of treasurer (Edward M. Spencer, 3–1–57), of secretary (Anthony G. Maihofer, 5–1–58), of controller (Eldred H. Scott, 4–14–53), and of general accountant (E. E. Roll, 5–7–56).

There was an obvious effort in these appointments to reach for younger men. In 1952, the average age of the seven men who were about to become the new management team for Detroit Edison was forty-seven. This group would remain substantially intact for a decade and a half.[7]

The Management Council was a venerable and important institution, despite the informality which marked its sessions. Attendance was not required but the members were almost certain to be present except when absent from the city.

Even the membership was rather informally determined. The officers of the corporation were, of course, members, as were the various managers. The general order describing administration named the chairman of the board as the chairman of the council, and specifically indicated that the membership was at his designation. The president was vice chairman of the council.

The council was a body of convenience, rather than of authority. No collective decisions were arrived at, no votes taken and the informal minutes are chiefly for the absentees' information. The Council functioned much like a cabinet, except that its size tended to limit discussions and debate. (The *Table of Organization* for July, 1968, lists 28 members.)

When disagreement develops, or when departmental policies appear to be contradictory, the matter is almost certain to be lifted from the floor for decision elsewhere. The council is thus less a place where decision is made than where it is announced. This is an important role, however. In any great bureaucratic organization, rumor far outruns the printed word. The council provided the means by which full and accurate information could be exchanged between department leaders and the top administration, and thus be made available across the Company. The discussion might be on as trivial a matter as the abuse of a coffee break, or as important as an alteration in fringe benefits; the vital thing was to secure direct communication and complete understanding. Morale is a hard word to define, but certainly one essential element in it is confident expectation of fair and equitable treatment. The council provided the means by which these goals could be reached throughout the Company.

For many years, Walker Cisler was chairman and presiding officer at council meetings. At his direction, the members made their reports. The brief statement could include the condition of the operating system, construction programs, labor difficulties, financial matters, marketing conditions, problems in regulation, or with government agencies. The chairman and the members could ask for supplemental or explanatory facts, when the topic seemed to justify or require exploration. Usually the report included the statement of decisions already made or actions taken.

Cisler's great skill lay in the quiet but effective way in which he merged pride in departmental competence into confidence and pride in the Company. Detroit Edison is not by tradition a modest Company, even in its private posture; the men who led it shared a faith in the Company and in each other.

The practice of the Company included extensive use of group or committee action. The *Table of Organization* for 1968, for example, lists twenty-seven such permanent committees with interdepartmental responsibility and membership. These were not committees of the council, and they did not report there, but to a large extent they were made up of council members.

On occasion, responsibility in the Company could become a very personal thing. For many years, at nights, on holidays or on weekends, when the normal supervisory personnel were away, one member of the council was designated as duty officer. Formerly it was enough that he be available for communication, but more recently their physical presence was believed necessary in case an emergency called for critical judgments and, if necessary, to commit the Company to a decisive course of action.

The accepted pattern of employment in Detroit Edison was consistent. There was a rather high turnover among new employes. But after five years, separations were few and Edison had become a career. The result was that by the decade of the fifties, long tenure in employment was the rule rather than the exception. The Company was grateful for such loyal and continuing service, but it also raised several problems, some of them serious.

The able young team of management which Walker Cisler picked to run the Company were troubled by the fact that they had subordinates who were senior in years and tenure, and who had been passed over in the creation of the new regime. For such persons, further promotion into the ranks of second level management was most unlikely. In some degree, seniority had become a handicap for ambitious men within the Company.

The fact was that too many Edison men had grown old together;

a department or a technical service might have its top men all in the same age bracket. There were roughly nine hundred engineers, scientists, lawyers, and others of specialized professional training in the Company. During the war decade few could be added and in the decade of the fifties, the declining number of men on the payroll made recruitment unlikely; the advancing average age was obvious, and with it, the inevitable clustering of retirement dates.

In 1960, after a detailed investigation, Harvey Wagner made a report which underlined the problem. Wagner was assistant vice president and manager of system development; the report on manpower is only an extension of his concern for the future, and involved especially technical and specialized personnel. The report forecast the loss of half of the key men in the Company over the decade beginning in 1960.[8]

The obvious thing to recommend was the recruitment of competent young men, but this alone was not enough. Realistically, Wagner associated with this a program of development within the Company designed to establish the potential of these recruits within five years. He recommended further efforts to encourage more rapid growth and progress for qualified men already in the organization.[9]

With the problem thus recognized, it became the responsibility of Sylvester Leahy, vice president for employe relations, to develop and implement the necessary program. He and his group made a chart which scheduled by departments and offices the dates for retirements, and thus dramatized the need to schedule replacement and allow ample time for training.

Leahy reported to the Business Club, an employe organization, on the meaning of this sequence of management decisions. Essential positions, now occupied by some older men, must be vacated to create opportunities and learning situations for their replacements. On a rather generous scale, premature retirements were negotiated with a number of the older employes in order to clear the road. This was tactfully referred to as a Special Separation Allowance. Oddly enough, Chet Ogden found himself forced to defend this Company policy when, in due time, the allowance system was terminated and the conventional retirement requirements reestablished.

The success of this forward-looking program would depend on

recruitment, and especially on the ability to reach engineers and others with the technical skills newly demanded. Recruitment programs were instituted in schools and colleges, and summertime employment was offered in the hope that permanent association would follow. Over the years, Detroit Edison has found the competition for men of promise increasingly vigorous, and has been forced to widen its areas of search.

Chapter Six

THE PROBLEM OF POWER PRODUCTION 1951-1958

War Experience and Company Planning / The Building of St. Clair / The Building of River Rouge / The Economic Measure of Success / St. Clair Extension and an Overview.

Before World War II, planning for power plant or other construction projects was a kind of group decision based on intuitive judgment. War time production needs and materials allocations demanded more formal action. Detroit, including the decentralized area which extended for fifty miles about it, and indeed all of Michigan was soon totally involved in war production. Edison officials estimated that eighty percent of the industrial load was going directly or indirectly to war.

The War Production Board in Washington was thus properly involved with the power resources of so vital an area. The board was concerned about the reserve margin for one thing, and also about the capacity of the electric power plants to accept new loads on schedule. D. J. DeBoer, chief power engineer for the War Production Board, inquired into the Detroit situation and began an extended correspondence with S. M. Dean of Detroit Edison which ran through the war.[1]

For most of Michigan's lower peninsula, five significant sources of power were involved. To the west were Consumers Power and the city plant of Lansing, while in the east were Detroit Edison, the municipal plant of Detroit, and Ford Motor Company's privately owned generating situation in Dearborn. There was pressing need, as war production mounted, to make any one producer's surplus power available to aid any other, especially in the event of emergency. This was not a new idea in the industry. The Detroit Edison Company with its four plants was obviously familiar with the idea and indeed it already had in use interconnections to Ford, to the plant of the Public Lighting Commission and most importantly, since 1927, to Consumers Power.[2]

In the war situation, however, the demand for production at high level and around the clock made difficult the customary shut down of individual machines for inspection and maintenance. The local committee under Dean, at the suggestion of DeBoer, began a monthly reporting system by which such outages were scheduled for the whole interconnected system. They could not anticipate accidents or breakdowns but they could contrive a kind of pooled reserve against such a contingency.

As the war dragged on, the War Production Board asked the same committee for an estimate of anticipated load for one or two years in advance. Again, the purpose was obvious, but it forced the Company to become formal in handling what to that moment seems to have been much less exact.

Walker Cisler, on his return to the Company, brought from his extensive war activity precisely this same kind of experience. His Washington assignment had allowed or even imposed a concern with inter-company functioning. The War Production Board was less concerned with the name of any one company than with its capacity to supply the vitally needed service to the nation. In Europe, also, and especially in his function as advisor to postwar governments, Cisler had again been concerned with an overall regional capability, the capacity to get the job done.

After the war, the Federal Power Commission continued to request information from the same producers. The information required was not quite the same but it continued to emphasize the idea of forecast and potential load. The Federal Power Commission did not have jurisdiction over the whole industry, and was resented by some parts of it. Perhaps for this reason the Edison Electric Institute (EEI), a membership organization, began collecting the same data. The report from Detroit Edison, September 30, 1948, signed by Walker Cisler, was sent in duplicate to the two organizations, the EEI and the FPC. It is interesting, and ultimately important, that the report went to an EEI national committee, whose chairman was Walker Cisler.

As chairman, Cisler had asked the president of Edison Electric Institute to request from members of that private organization a duplicate of the Federal Power Commission Form 12 E.[4] The response was generous and the reception of his report such that the informal first reports promptly gave place to a more formal style. In October, 1949, the Institute issued what it called the Sixth Semi-Annual Report, though to get five issues in 1947 and 1948, some very informal preliminary documents had to be postdated.

The reports of the EEI were, in fact, prepared at Cisler's direction, in the offices of the Company itself, and they continued to be through the years following.[5] It was a summary of current national capability, immediate expectations and longer prospects, both firm and tentative.

This concern with facts and prospects naturally rubbed off on the Company itself. The earliest "load forecast" for the Company available in document is dated May 11, 1948, and is presented not as the work of a committee, but as the result of a "Conference" which included Samuel M. Dean, William Hurley, Carl Mueller, Harold Reasoner, and Howard Seelye, who drafted and signed the report. The conference illustrates the informal processes which had been in use. The following winter the group was formalized as a committee whose first report was dated December 19, 1948. With no reference to any earlier report, this one noted the various classifications of service, took recognition of the factor of diversity and then estimated the peak demand for each of the next five years.

Over the years which followed, the Load Forecasting Committee changed somewhat, though the purpose remained constant. The first reports were exceedingly simple, making a mere arithmetical projection of a current growth rate, with a highly subjective adjustment of the projection to the expected economic climate. About 1956, the report became much more elaborate in content, in form of presentation, and in the statistical inclusion of national forecasts.

The problem which the president faced, and which this committee was especially directed to consider, was the matching of construction and increased capability to the demand which the Company could anticipate. From 1948 through 1955, the forecast had to be repeatedly, even annually, revised upward. Despite the fact that the production doubled, the peak annually pressed dangerously close to capability.

The past, with its cyclical swings in the economy, continued to influence the prognostications. Despite the showing for the current year, and maybe next year's favorable prospects, the committee felt the Company must not overbuild, that an economic decline would bring relief from continuing pressure on capability. This was caution, for recession was a part of experience, but year after year the increase in demand ran ahead of expectations and thus ahead of plans.

As president of the Company, in 1951, Cisler inherited the engineering decisions he had helped make. Originally, the plan had been to make the kind of extension-substitution at Marysville that had occurred at other plants, but the Korean emergency suggested that the Company would soon need a large block of new capacity. Two machines at Marysville would require the preliminary elimination of three small units there and that seemed unwise.

As early as 1949, it was known within the Company that plans called not for one but for two new plants.[6] This was an exciting announcement. A new plant, on a site free from the restrictions of existing structures, would be a challenge to the engineers. The last totally new plant had been built in the early twenties at Trenton Channel.[7] The effort to utilize there the newest and most efficient among the tested methods had been an outstanding success. It was twenty seven years since that achievement and the men of the Company were eager to try it again.

The first of the new plants was planned for the river front site near Marine City, some fifty miles above Detroit and not far from the Marysville Plant. The original cost estimate was thirty-five million dollars, and the plant was to have two machines of 150,000 kw each. Made cautious by the continuing explosive growth of demand, both output and peak, the Company had taken an option on a second pair of machines. Piles for the first unit were still being driven when the decision was made to contract for all four. January 1, 1953, was the target date, but in fact both of the first

The turbine generator used in Delray #2 (1909) resembled the vertical type, smaller machines used in Delray #1 (1903). This machine was rated at 14,000 kw.

The second floor of the Washington Boulevard power station, showing the Edison direct current, by-polar generator.

The turbine room of Delray #3 shows the type of generating units installed 1920-40. These are rated at 50,000 and 75,000 kw.

River Rouge Power Plant was new in appearance and in equipment; the three stacks served three boilers and the three turbine generators. 1956-7-8.

The St. Clair plants became operative in 1954. The two extensions are from 1959-60 and from 1969.

St. Clair #7, operative in 1969, is rated at 500 OOO kw.

units ordered were on the line before that time. The other two came on in June and October, 1954.

As they stood in the turbine room, #1 and #4 machines were from General Electric and #2 and #3 were from Allis-Chalmers. In both the power house itself, and in the individual machines, the engineers had taken full advantage of the freedom which the new site provided and several innovations were present. This was the first experience with the reheat cycle and the turbine-boiler unit. The high pressure boiler produced steam which was superheated to 1,000 degrees and to a pressure of 1,800 pounds per square inch. This steam was discharged from the turbine, at 450 pounds and 675 degrees, and was immediately reheated in the boiler to 1,000 degrees. At this pressure and temperature, it went to the intermediate or low pressure unit, which appeared to be a totally separate turbine generator standing beside the high pressure unit. The high pressure unit ran at 3,600 rpm to produce 40,000 kw. The low pressure unit operated at 1,800 rpm and produced 116,000 kw. The Company rated the four machines at 625,000 kw.

The building was a sharp divergence from the earlier Edison plants. The stacks stood at ground level instead of on top of the boiler house. Other heavy equipment, including fans, also was brought down to or near grade level. The significant lightening of the weight made possible considerable savings in structural steel, and promised easier and more convenient maintenance and upkeep.

A pump and screen house stood directly by the river. Instead of the conventional intake and overflow canals, steel and concrete tubes ran to and from the river. Since the plant burned coal, the location along the river's edge gave economic access to the barges and coal freighters.[8]

With a bow to the traditionally recognized father of the industry, Thomas A. Edison, the Company linked the formal opening of the plant to the October Diamond Jubilee of Light being celebrated nationally. A second coincidence was also noted. The completion of this plant brought the capacity of the electrical industry of the nation above one hundred million kilowatts for the first time. The nation's capability had increased by 11.5 million kilowatts in 1954, larger by a million kilowatts than any earlier year's increase.

Some time during the year 1954, the Company added its one millionth domestic customer, though which one he was there was no way of finding out. In 1953 and 1954, the annual production in kilowatt hours exceeded ten billion for the first time. This gave Edison 2.4 percent of the production and sale of the nation as a whole, and made it, according to the Company, the fifth largest in the nation—though there might be some quibble over what was being measured for size. Both the Edison Company and the industry nationally were making spectacular growth.

By now, the Company was in the process of creating its second "new plant," this one near the mouth of the River Rouge from which it took its name. Piling for this construction, a difficult and extensive preliminary operation, was being driven while the St. Clair plant was being completed. Even more than at St. Clair, the engineers reached for new and rather spectacular accomplishments.

For the first time the power plant was built without a basement. The pilings were cut at ground level and a concrete mat laid. Above this a space three and one half feet high was filled with sand, and the wires and pipes of the normal basement were placed here. Above this there was a concrete slab a foot thick, which became the floor of the power house.

At River Rouge, as at St. Clair, the stacks rested on the ground, together with such other heavy equipment as could be put there. The boilers, one for each machine, were the equivalent of fifteen stories high in interior dimension and the condensers would borrow from the river as much water as was used by the City of Detroit.

The exterior of the plant showed clearly its form and purpose, and no effort was made to give artificial beautification. The exterior walls were metallic, with pre-attached insulation for speedy and economical construction. It was frankly a machine, designed to utility and built to stand in a world of work. To many, the angular functional approach proved to be artistically successful as well.

The plan called for the installation of two machines, though space was reserved for another. The first two were 260,000 kw each, proudly hailed in Company publications at that time as the world's largest. Each, it was said, would provide power for a city the size of Cincinnati.

The first of these new machines, River Rouge #1, was completed in February, 1956.[9] The second, which brought the capability of the system to a new high, came on the line on November 18, 1957, and that evening a new peak demand occurred. Without the new machine, the margin of reserve would have been extremely close. By this time unit #3 had long since been ordered, and in fact went on the line only two months later.

The builders took advantage of the fact that the stacks stood on the ground and introduced a type new to the Company for #3. This stack was of poured concrete and the flag which flew over the completed structure was 425 feet in the air. The forms for this concrete structure were built so that they could be reduced as the stack rose, to give a pleasing tapered effect. The cost was about thirty-five percent less than for a conventional stack of steel and brick.[10]

Prior to the construction of the River Rouge and St. Clair plants, Detroit Edison had installed no units exceeding 100 kilowatt capacity. Units #1–4 at St. Clair were 156,250 kw capacity. The rule of thumb held that no single machine should be above five percent of the Company's capacity, because of the crisis which would be created by the failure of such a machine. By the time the two 260,000 kw machines were ready for duty, that system capacity was large enough to justify their size. The third machine at River Rouge, with a capability of 321,000 kw, was the largest machine on Company lines to date.

The two plants, and especially the new unit at Rouge, justified every hope. Expenditures for construction had been highly satisfactory when measured in the conventional terms of cost per kilowatt of capacity. Construction costs in 1915, 1927 and 1956 stood at $80, $90, and $135 per kilowatt. Measured in terms of 1937 dollars, the new plants were a bargain; Rouge on that basis cost roughly sixty percent of the cost per kilowatt of 1917.

The real measure of the success of a plant, however, is the cost at which it can deliver power. A great deal of the history of the Company and the industry, as instruments of the new technology, is revealed in such statistics. The old plant at Delray had used 23, 470 Btu per kwhr. The plant at Trenton Channel could produce one kwhr for 14,470 Btu. Given the coal costs of 1929, the difference then was between .32 cents per kwhr and .20 cents per kwhr, and that difference justified the destruction of the Delray plant in which millions of dollars had been invested.

The first two units of River Rouge, each 260,000, used only 8,900 Btu per kwhr, and the huge third unit used 8,800 Btu. This was roughly eight-tenths of a pound of coal per kwhr. At Delray 1, the cost had been two pounds of coal and Conners Creek had used one and one-half pound of coal per kwhr.

These figures show without doubt that small improvements in efficiency, because of the volume of coal used, can add up to significant savings. Despite the rising cost of living in the postwar world, the electrical industry has managed to reduce the average cost per kwhr consistently. The economics of scale, which technological progress made feasible, are most significant.

On expansion as rapid as this and on machines of this size, commitments were of necessity made several years ahead. In 1956, for example, when Rouge was being finished and its three machines were not yet in place, commitments had been made to add units #5 and #6 at St. Clair. These five additional machines, over a five year period, 1956–60, added 1,480,000 kw to the system capability. The increase was larger than the total capacity of the Company in World War II. When St. Clair received #6 in 1961, it became probably the largest thermal power plant in the nation, and the generating capability of this one new plant exceeded that of the total system when Cisler assumed the presidency hardly a decade before.

CHAPTER SEVEN

RICHES TO RAGS AND BACK 1955 - 1965

The Depression Experience / The Recession of 1958 / Dividends and Capital Costs / Economy in Production and Maintenance / The Use of Manpower / Strengthened Intracompany Controls / Economic Recovery and Its Effect.

CRITICAL DECISIONS FOR THE COMPANY CONTINUED TO BE MADE BY THE board of directors, acting on the advice of Walker Cisler and the corps of associates and assistants he had developed. The experience they shared in the Company and the knowledge they had of the economy, combined with temperament to shape and influence that advice.

Cisler's own experience was of vital importance. he was old enough to have known two wars and a depression, and to have seen the wild swings of the economy over almost a half century. In the Company itself, he had shared with Jim Parker in the hectic struggle to meet the surging industrial demands of postwar Detroit. He had been intimately involved in that decision making, and in dealing with the consequences of decisions thus made.

The Company to which he had returned after the war was not insignificant. Its assets at that time put it among the top one hundred largest corporations in the land and inside the largest half dozen of private electric utilities. But as large as that organization had been when he joined it, Cisler and his contemporaries had seen that growth of a half century equalled in a span of eight postwar years.

On the basis of this most recent record, the Company planners in 1955 projected a comparable growth rate for the next five years. The building program was calculated to meet the rising demand to 1960 and beyond, even at the high current rate of expansion. (Chapter Six) There was no alternative to such a decision, given the record.

On the other hand, statistical charts offered no guarantees that either the trends or the pace would continue. Inside the Company, because of the serious penalty imposed by excessive investment and unused equipment, the possibility of a depression colored every prophecy. It was felt that such an economic decline could be expected, and that it would relieve the mounting pressure on the Company's productive capability.

Despite such fears, year by year the demand had run ahead of expectations and thus of plans. The year 1957, with the River Rouge plant in production, was the first in a decade in which the capability of the Company was safely above the peak demand.

In part, this was the result of a kind of stabilization of demand. From 1955 to 1957, inclusive, both the peak and the total sales seemed to hit a kind of plateau. At first this was a relief, but rather quickly it proved to be something more than normal. Late in 1957, and even more so in 1958, the long expected "depression" was a reality.

The effect of the economic decline was, of course, felt throughout the community. The *Index of Industrial Activity,* prepared by the statistical division of Michigan Bell Telephone, shows a sharp drop in 1956 to a low in 1958, from which point a recovery began. This index, in which automobile production plus gross employment constitute forty percent, is accurate enough for present purposes.

The Company's experience paralleled this index, except that there was a little time lag in the depression's effect on electrical demand. Normal demand extended into late 1957; 1958 showed a very sharp drop, and the 1959 recovery was hardly more than a return to the level of 1957.

The effect of the economic slump showed in the records of Company income. Industrial revenues, which constituted roughly one-third of the total, were substantially fixed in the period 1955–58 inclusive. The year 1958 was the first peace year since the depression in which the total revenue of the Company was less than in the preceding year.

Even the sharp climb of the December peak slowed. As this occurred, and the new machines came on the line, the margin of safety became larger and larger. The peak load for 1960 exceeded that of 1955 by an amount which was hardly larger than the capacity of one of these new machines, and in these years, four had come on the line. There was a touch of irony in the situation which found the Company at long last lacking in customers rather than in capacity.

The 1958 depression was felt across the Company. The year 1957 had been in large part a prosperous one for the Company despite some warnings late in the year of troubles ahead. Earnings per share were $2.62, the highest in Company history, and for the seventh consecutive year, the shareholders received an increase in dividends. The rate in 1957 was $2.00 per share, up from $1.20 in 1950.

At the end of 1958, there was no mistaking the reality of the depression, now almost a year old. The earnings dropped abruptly from $2.62 to $2.17, a decline which was exaggerated by the increase in shares outstanding. The dividend at $2.00 was maintained, but earnings narrowly covered this figure, and there was no talk of an annual increase. This dividend rate of fifty cents per quarter was maintained for several years; the payment ordered by the board on March 16, 1959 was noted as the 200th consecutive quarterly payment since 1909.

Company reaction to the new economic climate was almost automatic in some areas.[1] The building program, announced just at the end of 1955 for the period 1956–60, was under contract; the construction could, however, be delayed a little and the costs thus spread over another year to 1961. In 1960, both the budget committee and the systems development committee agreed that further construction of power plants should be delayed for some time.[2]

Between 1951 and 1957, inclusive, the construction of two new power plants, together with modernization and extension of other plants and lines, had created a constant demand for capital. Company practice was to borrow from a bank or insurance company, usually at attractive rates and then to finance these accumulated short term debts with bonds or stock. In the 1951-57 period of construction, the long term debt, including outstanding debentures, rose from $229 million to $483 million. This figure included the debenture and bond sales of 1956 and 1957. The stockholders equity rose in that same time only from $199 million to $334 million, and thus stood at an unprecedented level of 41.3 percent at the end of 1957.

The impact of the depression tended to cut off plans for further extensive capital investments, and to slow down the pace of expenditures committed.[3] For several years, the total capital needs were met with internally generated funds. The conversion of debentures into stock, together with scheduled payments on debt, served to reduce the long term debt and by 1963 the traditionally accepted parity between debt and stockholders' equity was restored.

Company adjustment to depression called for a good deal more than a mere delay in contracted construction. The largest single economy came from the use of the new machines themselves. In 1958, when things were about at their worst, the six new machines at Rouge River and St. Clair produced a steady 1,150,000 kw, and carried the base load of the Company. Marysville was almost shut down, and other plants operated often only as standby and peaking service. The new machines required only about 75% as much coal per kwhr as did the older equipment. In 1961, when St. Clair #6 came on the line, the system heat rate was further reduced from 10,060 to 9,740 Btu per kwhr; this seemingly slight reduction of 310 Btu meant a saving to the Company of $1.25 million per year. Including the new machine, River Rouge and St. Clair produced 84% of the generated output in 1961.

Of particular importance was the development in the field of maintenance. In 1954, a rather important alteration in the Company's internal organization had occurred. A department of Construction and Maintenance—which shortly became "C and M"—was created. George Porter, vice president, directed the development of the plans and the organization, and the department reported to him.

The new department obviously did not represent a new function in the Company. In the past, with the small and relatively simple machines and equipment, even major repair and maintenance had been the responsibility of the crew assigned to each specific plant. The new arrangement left the plant in the hands of only the operating force, permanently assigned. A mobile crew from C and M would be sent to the plant for major construction and maintenance work.

The new department rapidly developed the skilled manpower essential for its important and varied assignments. The force differed in size, though normally it was seven to nine hundred men. Methods and equipment also had to be developed for the department, and this was a field in which Cisler had personal experience and a special professional interest. In 1960, he summed up the recent experience of the Company in an address to the American Society of Mechanical Engineers meeting at St. Louis.[4]

In this address, Cisler used the term Maintenance Engineering to embrace new Company practices, especially as they had been developed since 1957. It referred to a systematic application of preventive maintenance, which the new department made possible. The traditional system had called for maintenance operations on a fixed time schedule; thus, a turbine generator would be torn down for examination and such adjustment and repair as it required every three years. The purpose of "engineered maintenance" was to substitute another system of schedules. Within the Company, and with the cooperation of equipment manufacturers, the department sought to develop tests and instruments to discover malfunction, and to check efficiency as a continuing part of operation. They sought to avoid taking down machines which were functioning well, and to discover those in need of adjustment or attention without regard to time lapse.

The electrical industry had long had a horror of outages, and tended to over-react in prevention. In that complicated complex which was a power plant, the failure of a pump, for illustration, might jeopardize operations or even the plant itself. The temptation was to insure continuity by installing a spare part, even when the part was itself an expensive machine. Now it was discovered, periodic routine testing could check functioning and give assurances of reasonable security without the extra capital cost of duplicate installation.

An attempt was made to reduce the meaning of these and similar practices, and the overall efficiency of the new maintenance system, to dollar costs. In 1952, before the creation of the new department and the development of its methods, the cost annually to maintain one megawatt of capability was 825 man-hours, each costing $2.25. The total cost to the Company was $5.7 million. In 1961, with labor at $3.89, at the old hour requirement the cost would have been $13 million. Instead, it was $6 million. The number of men in C and M had been cut from 1,595 to 954, though the system's capability had almost doubled.[5]

The stress of the difficult times reinforced and emphasized the standing efforts toward economy and efficiency in the use of man power, although this was a possible economy which would have to be practiced with care. From 1951 and 1952, which represented the high point in the number of operating employes, the management's attention to this problem brought about an annual reduction. At first the reduction followed normal retirement of persons who were either not replaced or were replaced with employes who could be transferred.

The large machines with automatic operations extended and improved, demanded fewer operators despite their much greater capacity. Of course, such manpower replacement was not limited to the power plants. In modern technology, the substitution of the machine for human labor has been a common place occurrence, happily associated with the rising wage rate. In the Company, the reduction of employe numbers did not balance the rising wage level, but tended to hold that increase to a level which the Company would find tolerable.

The reduction in numbers of employes continued, with the pace stepped up when economic pressures seemed to justify it. The compensated early retirement system (Chapter Five) provided a means by which both to ease and speed the process of separation. Between 1956 and the end of 1959, the total number of employes had been reduced almost 10 percent; when the so-called operating employes alone are considered, the percentage was slightly larger. In 1950, Edison had employed one person for each 110 customers; at the end of 1960, the ratio was one for 171.[6]

The board was informed that at the close of the year 1960 the employes numbered 9,921, a reduction since January, 1957, of

1,267. This figure includes both operating employes and those assigned to construction.

In moments of restricted activity, when the economic situation demanded both economy and efficiency, it was especially important that balance be maintained within the Company. The Company was served by groups of highly dedicated experts whose concern with specific segments of the Company could inhibit understanding of the Company as a whole. It was essential that the central administration develop a method by which decisions could be coordinated.

Three committees named will serve to illustrate the problem, and to suggest the methods by which the management attempted to deal with it. It is not suggested that the concern for costs was new. The modification of the organization indicates only the formal recognition of a problem with which management had long been concerned.

The load forecast committee could make only a limited contribution to the Company, for everything depended on the acceptance and use of the information by leaders of the several departments concerned. When or if they substituted departmental judgment, then to that degree coordination would suffer.

A new department of Systems Development was created in 1958, with Harvey Wagner as manager. In the same year a new committee on systems development was organized and charged with overall responsibility. The committee with Harvey Wagner as chairman brought together members of the older forecast committee, whose work continued, and also key persons from the several departments.

Wagner was thus made responsible for reviewing all proposals for construction and expansion and, with his committee, issued what was substantially a preview of the year's approved activity. The first of these elaborate reports, for the year 1960, was issued in December of the preceding year. Fundamentally, this was a translation of the annual budget into a plan of action.

But construction of power plant facilities and transmission lines is not a matter of one year. The projection of load demand for the next five years was restated in terms of project construction; obviously, the firmness of the estimate for the ensuing year could not be matched over the remaining four, but the timing of operations could be estimated and essential consistency maintained in departmental plans.

A second committee, with some interlocking membership, was on Capital Expenditures, with Eldred H. Scott, as chairman. This group had the responsibility of translating the plans for construction and development into an implied financial commitment, and then of working out the financing involved. Here again the budget figures for the year ahead could be quite exact, but accuracy declined through the projection over five years. In legal status, the Capital Expenditures Committee was advisory to the controller, whose approval was essential on each work order.[7]

The third committee had to do with the use of electronic data processing, at which the Company had already had some experience. The problem of customer accounts, both in billing and collecting, was a major one. In 1953, in collaboration with IBM, Howard Stevenson from sales and J. Douglas Elliott from accounting, with the approval of A. S. Albright, then executive vice president, began work on the problem. In October, 1954, they submitted a joint proposal, chiefly related to customer accounting, but rather quickly it became clear that the techniques had other applications.[8]

The economic situation in 1958 suggested the immediate search for additional economies in office procedures, and a new interdepartmental committee was named. This third committee was called the Administrative Systems Advisory Committee; it superseded an earlier committee with more restricted responsibility. The chairman was Donald F. Kigar, executive vice president, and its membership included eight other vice presidents and managers directly concerned with business procedures, as opposed to production and distribution. The secretary of the committee was J. Douglas Elliott.[9]

The creation of the committee was followed at once by the establishment of a department on Administrative Systems Planning, and Elliott was made director of the new department. Its responsibility grew out of the increasing volume and complexity of the Company's office systems and procedures, and its function was to provide central coordination for these activities. It was to be the planning agency which would aid Company groups in gearing their systems and procedures to changes in both manual and automated techniques. This increasingly important part of Company activity was again recognized in 1968 when Elliott was named to the newly appointed position of assistant vice president and manager of administrative and engineering systems.

In the annual report of the Company for 1958, there is reference to the unfortunately reduced demand, when the Company was at long last so fully prepared to serve more customers. It was the operating economy of the new machines which enabled the Company to survive the business slump with so little permanent damage.

The effect of the business recession showed itself quite clearly both in sales of power and in cash received. In the years 1955, 1956, and 1958, sales had been a little above eleven billion kwhr per year; in 1957, a banner year, sales were twelve and one-half billion. With the return to a more normal business climate in the recovery year of 1959, the Company's sales reached 13 billion kwhr and in 1960 and 1961 exceeded 14 billion kwhr. In revenues, the story is much the same. In 1957, the Company received just over $250 million and in 1955, 1956 and 1958, less than that amount. In 1961, revenues were in exess of $285 million and in 1960 they had been just below that figure.

This rate of recovery from the recession was satisfactory, perhaps, but was not exciting. The directors' letter for January 13, 1960, carried the staff estimate (obviously with the approval of the president) for the next five years. The completion of the current construction program would bring the production capacity to 4,198,000 in 1961, an amount which was believed adequate for at least the five years covered in the estimate, to 1965. In 1961, in his address to the Power Club, Cisler used these same figures. On an earlier occasion, he had remarked that if demand had caught up by 1965, he would be glad then to consider further expansion of capability.

The rate of recovery in production and in income might appear modest, but the management could take real pleasure in these increases. In his letter to the board dated December, 1960, Cisler reported that over the most recent seven years, the increase in earnings had averaged 9.6 percent on an increase in revenues of 5.6 percent. With equipment in place and with many basic expenses met, the Company was marketing what would otherwise have been surplus power.

The effect was remarkable. In 1960, the earnings per share had been at their historic high, in spite of a rather considerable increase in the number of shares outstanding. The return in the year 1961 was even higher. In anticipation of the final quarter's report, late in 1960, Landrigan was permitted to make the official recommendation that the increase in dividend rates be resumed.[10] He contrasted the return which had gone to the shareholder with the wage level and consumer benefits to argue the equity of the 5¢ per quarter increase he recommended. The pay-out for this quarter and as anticipated for the following year would be 77%. In December, 1962, the stock of Detroit Edison was split two for one, and the new stock was promptly quoted on the market at a healthy figure.

The rising sale of electricity inside the established capacity increased the income without proportionate increase in costs, improving earnings and dividends. This situation, together with the completion of a rather hectic building program, may be responsible for the note of complacency in the annual report and other Company statements.

The annual report for 1962 reported record breaking production and sales, and 1963 followed the same trend. Peak production records of earlier years were broken not once but a dozen times in each of these years and, for the first time since World War II, the 1963 summer peak exceeded that of the preceding December. One favorable result was a gain in load factor, the highest in the Company history since the straining war years.

The increase in industrial demand was particularly noticeable. There were brief interruptions among customers as a result of their labor difficulties, but new industries continued to become new customers, and old ones prospered.

Beginning with 1961, the city and the country enjoyed successive years of uninterrupted prosperity. Industrial sales were up over the preceding year by 9.1 percent in 1962, 8.8 percent in 1963 and 9.4 in 1964. The increases in industrial sales thus confirmed increases in the more stable commercial and domestic load. But this kind of economic acceleration could upset any plans.

In 1955, when the first of the new and impressive machines at River Rouge plant went on the line, the torturing improvisation demanded by the December peak seemed to be ended. In the ensuing decade the problem had been surplus capacity. Now, suddenly, the old days of shortage had returned, and the yearly peak was again perilously close to the capacity. The demand which had been anticipated for 1967 was attained by 1965.

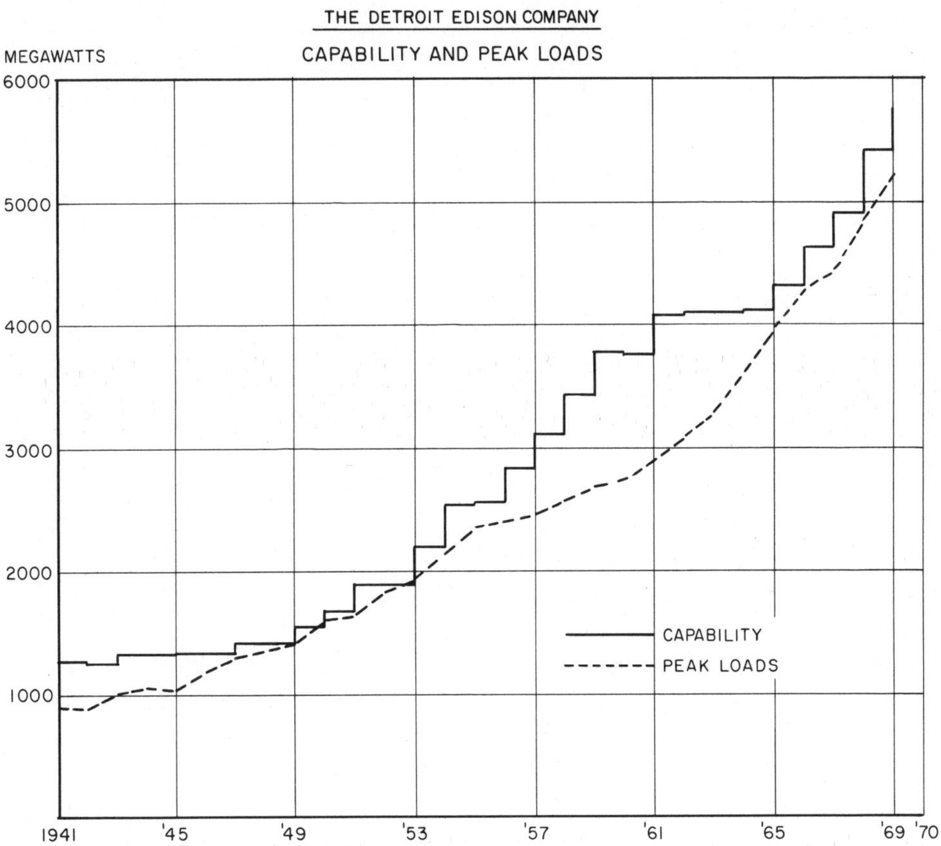

CHAPTER EIGHT

LINES AND SYSTEMS

The Physical Form of the Company / The Loop System and the Tower Lines / Population Drift: The Adjustment in Distribution Voltages / Transmission and the Extra High Voltages / Maintenance of Reliability and Service / Doing Something About the Weather / Plans for Disaster.

THERE IS A MYSTERY AND AN EXCITEMENT ABOUT THE POWER PLANT. THE cathedral like vistas, the awesome might and majesty of the flaming furnace, and, above all, the turbine generator room, where almost unbelievable power is marshalled in dramatic orderliness. It is easy to be romantic about a power plant, where men seem to serve machines, and where machines themselves seem to acquire personality characteristics—good and bad.

The men who build and maintain the transmission network also have their claim to drama and adventure. The whole area of distribution has its own technology, its own fundamental limitations, and its own story. The investment required to transmit and distribute electricity exceeds the investment required to produce it; the ratio is roughly five to four in capital cost. The physical form of the Company or of any electrical utility is determined by the transmission and distribution systems necessary to reach and serve the market. At its first plant location in Delray, Edison had moved to 22,000 volts for its cable distribution; this proved to be so near the top tolerance of the existing cables as to constitute a hazard. Even at this level, the transmission costs were high, both in capital expenditure and in line loss. With the expansion of the service area, a second power plant was indicated. The decision was made in 1911 to build at Conners Creek, on a new site, for reasons of expected economy.

The electrical system associated with these first two plants has been most conveniently compared to the forking and branching of a tree or, in this case, of two trees. The base was the power station, producing electricity for transmission at 24,000 volts.[1] At a so-called substation, this voltage was reduced for neighborhood distribution. At the familiar "tank" on an Edison pole, this voltage was again cut from 4,800 to 240/120 volts for domestic use in one or more houses. The analogy of tree and branches emphasized the dependence of any given line on its single direct connection with the power station, except that in some special cases the touching tips of the two "trees" suggested the establishment of alternative connections as a security measure.

Given the technology in transmission of 1912, Conners Creek Power Plant was essential. By the same requirement, the Company seemed committed to the cost and inconvenience of a plant each eight miles along its river frontage for as far as the load would go.

The Company escaped this cost and inconvenience only because of a drastic reorientation of the whole system early in the twenties. At that time, the Company built its first steel tower lines, designed to operate at 120,000 volts. The pattern established then was substantially still maintained in 1946.

The two new anchor stations were at the end of the loop extending from Marysville (1922) to Trenton Channel (1929) and reaching inland as far as Pontiac and Ypsilanti. Using step-up transformers, the power poured onto the high line at the voltage level new to the Company. At first, at three potential centers of demand, step-down stations reduced the voltage from 120,000 to 24,000 volts. Company terminology called these elaborate installations "stations" for they served exactly the same purpose as did the earlier power stations at Conners Creek and Delray. In later years the number of these step-down stations on the line was increased.

The loop, with its two power plants, introduced a transmission system for which a woven basket is an apt illustration, rather than the tree. For one hundred miles from one station to the other, the tower line was a kind of supporting ring for the system. From the stations along the river front, but also from the stations on the tower line in the interior, power fed toward the center of the "basket" and provided the security of service which came from multiple power sources. The four power plants of the Company became a single operative unit.

Even before the first steel tower line was constructed, the transmission at 24,000 volts proved inadequate for some of the distances involved; by an ingenious adjustment first one and then other lines were shifted to 40,000 volts. A given line at the new higher voltage transmitted almost three times the power and with less line loss. When the Thumb area was taken over and its local plants abandoned, the limited demand there was supplied by a 40,000 volt transmission line of considerable length.

Inside the populous parts of the city, such tower structures were unattractive, costly and dangerous. Other than being expensive to build, tower lines required purchased right of way or easement. The solution was to attempt to go underground, an idea with which the Company was long familiar for lower voltages. Not until 1941, did the Company achieve a satisfactory solution to this particular problem.

The demonstration of a successful 120,000 volt cable for underground use, in part developed within the Company, solved the difficulty. One of the new step-down stations, located on this new underground line, was placed in the center of the city of Highland Park—a much more acceptable neighbor than a coal burning power plant of more conventional design would have been. The high voltage cable was subsequently used to tie together the older power stations along the river, and to give convenient access to the downtown Detroit area.[2]

The importance of this distributive system to the Company is obvious. Its importance to the region and its economy is even greater. The system allowed not only the creation of mass power at low cost, but permitted its economical and flexible utilization at any desired location in a great area. Without this availability, both the economic and the human ecology would have been quite different.

The suburbs adjacent to the city and the towns and cities further out all shared heavily in the population growth. The city of Detroit proper was totally ringed about, and had no room for expansion and attendant population growth. The destruction of housing for expressways, the elimination of some of the antiquated and substandard housing, together with the outward migration of the jobs themselves, resulted in an actual decline of the population in the city proper between 1950 and 1960. Metropolitan Detroit had become more than a statistical phrase.

The population of the 7,600 square miles of Detroit Edison territory before World War II was roughly two and one-half million persons. Either the expected postwar exodus did not occur, or the loss was soon recovered, for by 1947 the area's population was three million, and it exceeded the wartime maximum. In 1956, it was four million.

Roads and superhighways cut time and cost of transportation, and both residences and jobs were scattered over a three county area—in time over six counties. The Company estimate of population served climbed to 4.9 million in 1969.

By that date, the Company recognized sixty-one percent of its service area as "urban." What that means is that boundaries were substantially without meaning, and that the word Detroit, as an industrial and economic entity, had less and less relationship to the Detroit which was the political and geographic entity.

The impact of all this on the demands for power is obvious. Even more dramatic is the demand made on the distributive function of the Company in its efforts to serve the extensive and scattered new load.

With this explosive drift of population and industry outward, it was increasingly essential that there be maximum economies in substations, rights of way, and lines. The Company had a third of a billion dollars already invested in lines and systems.

The decision was made (1959) to change at least some of the distribution lines from the voltage level of 4,800 to 13,200 volts. This shift would allow each circuit to serve more customers, and hence reduce the number of circuits required. Congestion in the vicinity of substations would also be reduced.

By 1960, then, the transmission system utilized 120,000 volts or 40,000 volts, depending on demand. Distribution from the step-down stations to a substation was either at 40,000 volts or at 24,000 volts, again depending on the demand. Distribution from the substation was at the new 13,200 or at 4,800 volts, and across the city and in the territory beyond, wherever electric range, washer or dryer, air conditioner or other major appliance demanded, the Company lines supplied both 120–240 volts to the household customer.

Under this kind of expansive pressure, the simple basket-like pattern of the 1923 distributive system became inadequate. The original loop was supplemented by an inner loop, an inner inner loop, an outer loop and other construction designed to meet the demand growing both in area and requirements. By 1960, the steel towers stretched across six hundred miles of Michigan landscape, with most of them carrying two circuits. New construction was scheduled at the rate of forty miles a year. Another two hundred miles of 120,000 volt lines were carried by wood or metal poles. Two thousand miles of 40,000 volt lines were in use.

The critical problem was of the high voltage lines, under the pressure of unprecedented Company growth. In 1923, when first the 120,000 volt tower line was completed, a single circuit would have transmitted the entire production of three of the largest generators then being made. When Rouge was completed with its first three generators (1957), it required ten such circuits to take the load. St. Clair, in the process of becoming the world's largest, would be in the center of a territory which in time could be covered by a veritable umbrella of towers and lines. The cost of right of way and construction was constantly growing.[3]

Higher voltages were not unknown. Lines of 230,000 volts served several hydro plants where distance was a factor. The Hoover Dam-Los Angeles line was at the unusual 287,000 figure. Several years before, Sweden had broken through the older limits, to show that 380,000, or even 400,000 volts, were possible and economical under their circumstances. In this country, the American Gas and Electric Company (now named the American Electric Power Company) had established the first network type installation with a standardization at 345,000 volts.

As early as 1956, when the new lines were being built to handle

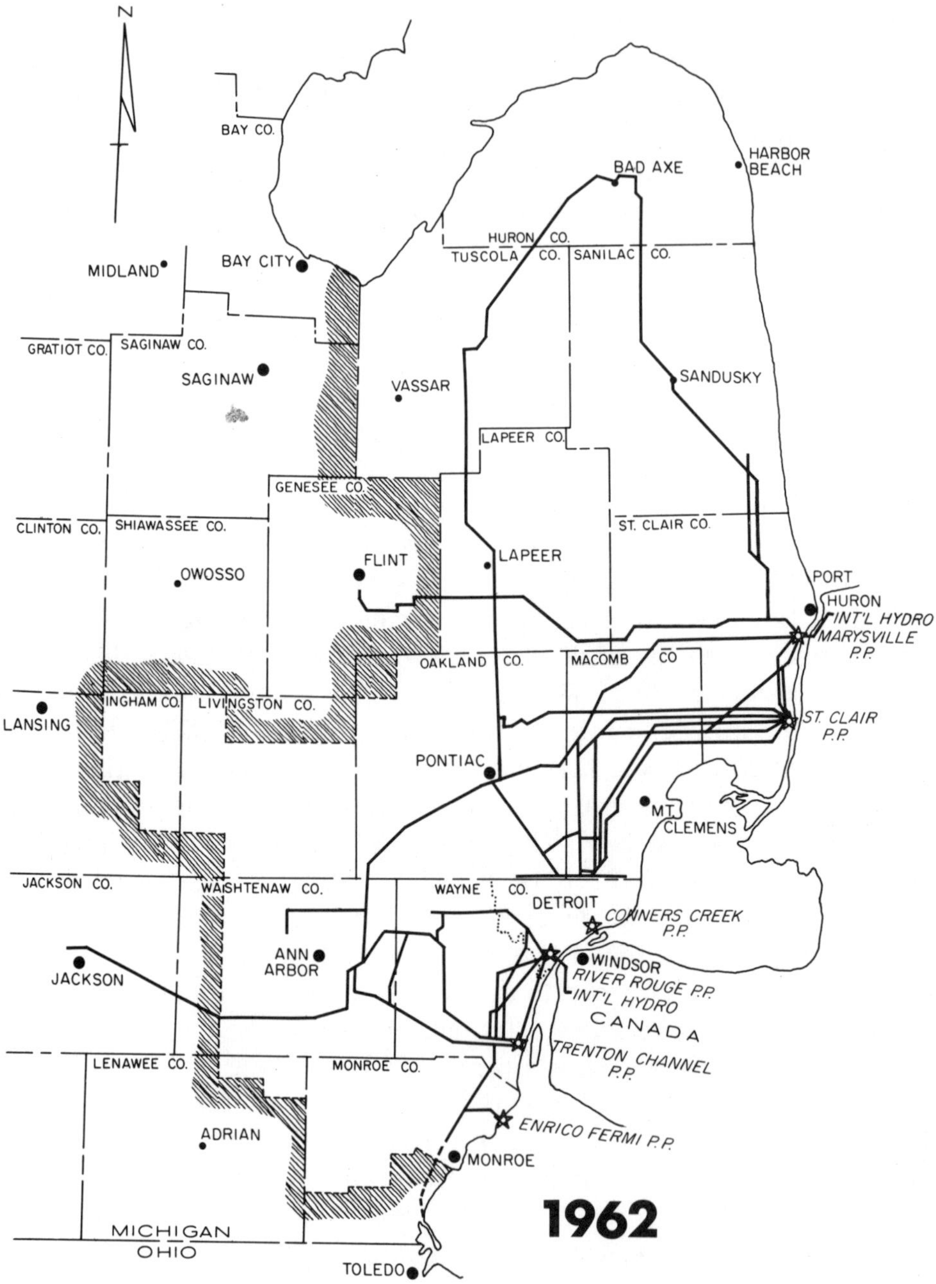
N
BAY CO.
HARBOR BEACH
BAD AXE
HURON CO.
TUSCOLA CO.
SANILAC CO.
MIDLAND
BAY CITY
GRATIOT CO.
SAGINAW CO.
SAGINAW
VASSAR
SANDUSKY
LAPEER CO.
GENESEE CO.
CLINTON CO.
SHIAWASSEE CO.
ST. CLAIR CO.
FLINT
LAPEER
OWOSSO
PORT HURON
INT'L HYDRO
MARYSVILLE P.P.
OAKLAND CO.
MACOMB CO.
INGHAM CO.
LIVINGSTON CO.
LANSING
ST. CLAIR P.P.
PONTIAC
MT. CLEMENS
JACKSON CO.
WASHTENAW CO.
WAYNE CO.
DETROIT
CONNERS CREEK P.P.
ANN ARBOR
WINDSOR
RIVER ROUGE P.P.
INT'L HYDRO
JACKSON
CANADA
TRENTON CHANNEL P.P.
LENAWEE CO.
MONROE CO.
ENRICO FERMI P.P.
ADRIAN
MONROE
1962
MICHIGAN
OHIO
TOLEDO

St. Clair #5, towers were designed to move to some higher voltage, though for the moment they conformed to the established 120,000 volts.

By the end of that year, the decision was made to use the 345,000 volt line. The Company's own research confirmed the acceptability of this level, and there would be some additional advantages if an industry-wide agreement developed on what was being referred to as extra high voltage.[4]

For the time being, no line was carried at extra high voltage, but continuing tower construction looked forward to such a change. About seventy-three miles of steel tower lines between St. Clair and Pontiac, begun in 1957, were for the moment carrying 120,000 volts, but would become the base for a totally new extra high voltage system by a mere change of the line and equipment.

The decision did not end all interest in ultra high voltage lines as others were developing them. Cisler, while in Russia for a meeting of the World Power Conference Executives (1961), visited South Station, the terminus of a line to a hydro plant 400 miles away; the line operated at 500,000 volts.[5]

Detroit Edison, together with some manufacturers and other utilities, supported research for several years at Cornell in the attempt to solve the problem of ultra high voltage cable for underground use. The cost of an underground line would be roughly seven to twenty times the cost of the overhead tower system, but there were circumstances in which the cost would be justified, if the technical problems could be solved.

All lines constitute a continuing threat to the Company's boast of reliability. The men in charge try to anticipate the unexpected, and be ready for it, but accidents do happen. On the overhead lines, an outage can be caused by birds or other animals, by a fallen tree or tree branch, by high wind or an ice storm, by a crossed wire, or even by faulty equipment. Underground lines are subject to "dig ins," can be affected by corrosion, and other unseen hazards—and are even more difficult and time consuming to repair. Even the steel tower lines could be victim to the bizarre. A train derailment with a refrigerator car smashing into a tower base could hardly be anticipated.

Preventive care was expensive but essential. Tree crews covered the lines every three years, and, with permission, cut back trees which threatened to cause trouble. Wooden poles were subject to rot, and the cost of replacement or failure justified an extensive research program in prevention.

For the big transmission lines, including especially the spreading steel tower lines, foot patrol gave way to inspection by air. At first a special slow moving airplane was used, but when helicopter service became available it was so satisfactory that the annual inspection by climbing the pole was no longer necessary.

Closing down a 120,000 volt line for inspection and repair was a serious inconvenience, to be avoided if at all possible. For a long time, the Company had had crews especially trained and equipped to make repairs and replacements on "hot lines" in service. This had been established Company practice for lines on lower voltages, but the techniques were expanded and after 1952, the Company trained its own hot line crews in a special school and included work at 120,000 volts.

The linemen are a group of some seven hundred men, normally engaged in routine construction and maintenance of the lines. They are divided into five so-called divisions, though the autonomy which might be implied by that term is limited. The size of the work force in a division differs, depending on the work load at hand. Routine construction work may be a part of this load but, frequently, perhaps usually, the available crews are not enough to maintain construction schedules, and outside contractors who supply their own work force are employed for a specific job.

The Company maintains some sixteen warehouses throughout the territory, where mobile equipment, material and supplies, and special tools are available. They are of course the centers for normal maintenance and repair operations of the district, but they are also prepared for emergency situations.

It is estimated that 75 percent of all outages are the result of lightning. For some reason, there are certain areas where such strokes are frequent, and there the service hazard is unusually high. The long lines of the rural service create a special risk area, for this and other reasons. Usually, however, service was better than the casual customer would remember. A careful study of outages and their causes is maintained, with a year by year summary prepared; the year 1964 is here chosen at random. Even in the rural areas and small towns (less than 200 persons per square mile), the interval between outages of as much as five minutes was

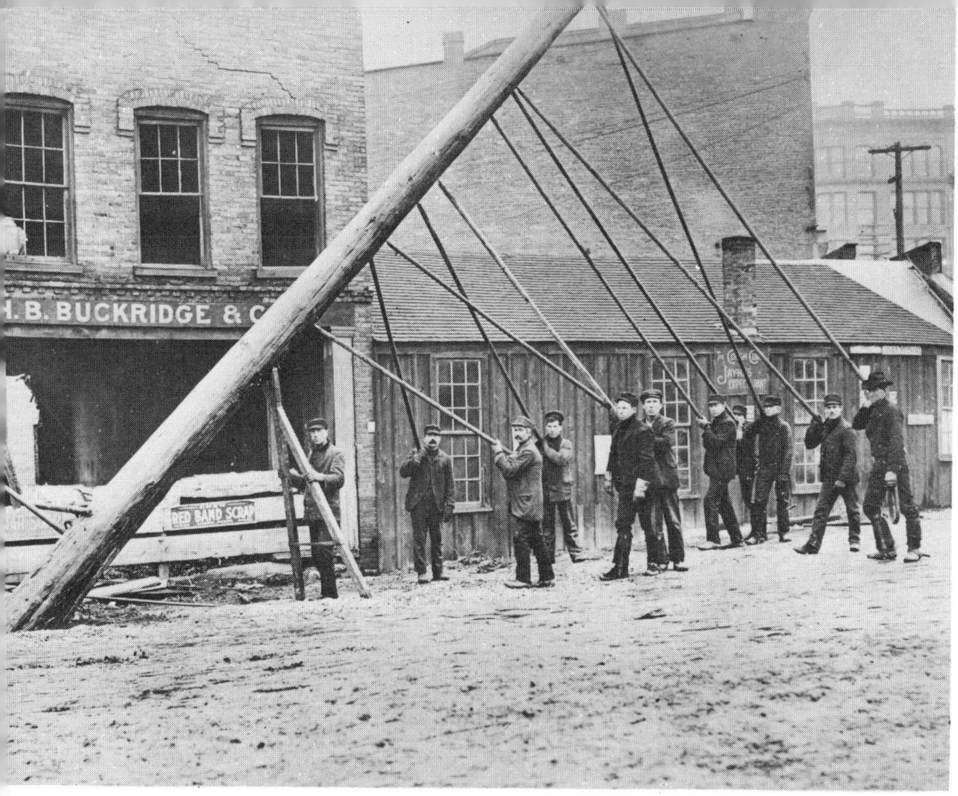

The construction once required human strength.

Steel power lines, at either 120,000 or 345,000 volts, required elaborate equipment for construction and maintenance.

twelve months, and between outages of an hour or more, the interval was two years. In the urban area (above 2,000 per mile) an outage of as much as five minutes averaged once in five years, and only once in ten years would the average outage be longer than an hour.[6]

Interferences with the operations of the system are much more frequent than most customers can know. When such an interruption occurs, automatically and almost instantaneously, the built-in safety devices are activated; with a flicker so slight the customer will hardly notice it, the damaged line is cut out, and bypassed.

Interruptions which come from a single simple failure of any part of the transmission or distribution system are almost always momentary, because of what is substantially a second or alternative system. Only rarely, and under very special circumstances, will both the line and the alternative be out of service at the same time; in such a case, the customer has no service until at least emergency repairs are completed. The cost of this system of alternative lines is quite large, but it is the price of reliability.

Weather becomes a matter of grave concern, and the Company has done something about it. For many years weather information from throughout the Company area moved quickly to a weather center, and was supplemented there by official Weather Bureau predictions. Through close association with other utilities on all sides, a kind of "early warning" of storms and winds as they sweep towards Edison territory is maintained. Within recent times, however, the whole weather service has become highly sophisticated, in large part because of the airlines' dependence on exact and comprehensive information.

Because of its concern with this problem, the responsibility for developing the Company's own elaborate weather forecasting service rested in the overhead lines department, where the major utilization would occur. This reporting service was important enough to be specifically commended in the annual report to stockholders for 1962. In April, 1964, Clifford J. Crockford received the Alex Dow award for outstanding service in the development of weather forecasting techniques for Company use.[7]

A constant flow of weather data from stations across the nation goes to Washington; there, the reports are coded and consolidated, and fed back to local weather forecasters and to such private users as the airlines and the utilities. The Company weather department selects from the teletyped material, that which is relevant to the local scene. The radar coverage from Detroit is roughly three hundred miles. To the west and south it is supplemented by reports from Ohio, Western Michigan and Chicago.

The warning of the impending storm, together with an estimate of its severity and extent, serves to alert the one or several divisions or districts in the area presumably to be affected. Before the storm hits, normally, the emergency crews are assigned posts in one of the warehouses, with motor equipment ready to move at the first indication of need. Sometimes the precaution is unnecessary, but more often the forecast is right.

Storm damage can be anything from trivial to extremely serious. Reports come from excited customers who cannot assess damage or even describe it accurately. Dispatching a truck and six man crew to an incident which may be trivial could waste resources seriously needed elsewhere. Across the Company some eight hundred persons from offices and other nontechnical work have volunteered for emergency duty at such times of crisis. They respond to calls on request, report back, and, in the event the damage is dangerous, they are prepared to stand guard until the repair crew can arrive.

A hurricane or other disaster of such proportions could create problems no utility could solve quickly. In some cases the damage could require a total rebuilding of the system. The custom of the industry, in such an emergency, allowed the disabled utility to borrow workers, equipment and supplies from one or several unaffected neighbors.

Edison had shared in this kind of service. In 1951, two crews went to New Jersey when a third of a million people lost electric power. In 1954, when hurricanes Edna and Carol ripped the New England area, Jack Drummond, then in charge of overhead lines, had crews organized by nightfall to leave by plane, and they were at work near Boston the next morning; four trucks and equipment followed by train. Before and since those dates, help has gone to Philadelphia, to Chicago, and, of course, to such nearby cities as Cleveland and Toledo. In referring to this type of aid, Cisler remarked, "in these cases, it is certainly better to give than to receive."

Fortunately, Detroit Edison's own experience with disasters of this proportion has been slight. In 1953, the tornado at Port Huron struck a limited area, although to the people of the stricken area it was not small. Damage was estimated at $2.5 million. The storm struck in the late afternoon. Within minutes the Company had fifty-six Overhead Line crews on the way. The streets were a tangle of downed wires and fallen trees, with no street actually usable until trees could be cut up and removed. Yet hospitals had emergency service in four and a half hours and most customers in twenty-four hours.

Experience in the New England area, and several lesser but important incidents locally, led in 1957 to the reactivation of a committee which had had long standing. Dating back to the two World Wars, it was first an agency of civil defense. In World War I, the concern was to be prepared for bombings; in later periods of national stress, especially after Detroit became so significant to national defense, the precautions extended to include sabotage, riots, and other civil disturbances. The essential purpose had been to develop a program of action to prevent or minimize such disasters, with the essential planning done well before the event.

Bill Williams, long an employe in responsible positions, had headed this wartime effort. Williams received the rather odd title of Disaster Director, and with a staff and committee he worked out an elaborate procedure and an instruction manual of formidable size. At least once, Williams and his group ran through an exercise of simulated defense. Fortunately, they had little other occasion for practice. The committee, as reactivated, was concerned with natural as well as man made disasters.

In the fall of 1968 a sudden windstorm hit erratically in the northern suburbs of Detroit, doing damage beyond any previously experienced by the Company. In this case fallen trees made streets unusable, and communications went out due to fallen phone lines —even television and radio were out of service. The situation was further complicated because this storm hit while linemen were on strike.

Later in the winter, a sudden ice storm swept the upper part of the Thumb area, followed soon after by a second and even more devastating freezing rain. Ice broke lines, toppled poles, felled trees and branches, and at the same time made roads almost impassable.[8]

In both cases, the loss of communication proved to be a most serious handicap to restoration operations, despite the fact that the Company had long had radio as a supplement to its interplant telephone system. The pre-planned mobilization system threw a thousand men into the damaged area, but directing their work proved difficult and delay resulted. The experience with the ice storm especially seemed to suggest the reexamination of the plans for handling such an emergency and a special committee from affected departments was asked to bring in a report. The committee included in its membership Paul A. Van der Meer, director of the Department of Civil Defense, whose regular assignment was closely related to this type of interruption of service.

Communications loomed as the major problem only if direction of all operations came from a single administrative center. The committee recommended a kind of "loose-link" system which delegated extensive responsibility for direction to the district office. Calls for assistance and directions for its use would come from the man or men at the scene who could estimate the extent and dangerousness of the damage, and whose knowledge of the local territory could thus best be brought to bear in the decisions on repair and priority. Committee plans called for a close cooperation between the representative of the sales and business offices in the districts and the local construction and maintenance leadership.

There were suggestions that this emphasis on local direction might be a forecast of a move not yet recognized toward a more decentralized administrative structure. The expansion of the lines and systems over an urban area so extensive as the one Edison served implied problems caused by variations in local conditions.

CHAPTER NINE

SALES AND THE DOMESTIC REVOLUTION

The Earlier Emancipation / The Meaning of Edison Service / The Organization of Sales / The Heating of Water and Space / The Domestic Customer in Company Affairs / The Company in Domestic Affairs / The Commercial Adaptation.

ELECTRICITY ONCE HAD BEEN A LUXURY, WITH THE PLEASANT YELLOW glow of the old carbon light the visible mark of an affluent household. In the urban regions, it had long since become a necessity commonly accepted. Living quarters which did not have electricity were hovels indeed. In the rural areas of the older Edison territory, electricity was everywhere available, and served all but a few farms. The decade after the purchase of the Thumb electrical service saw extensive construction in spite of the intervention of the war, and, even there, availability soon approached one hundred percent.

The decade of the twenties witnessed the first extended use of electricity which in a generation would constitute a domestic revolution. Small items such as the electric iron, the fan, the hair curler and even the vacuum sweeper may not have added much individually to the electric bill, but to the housewife they constituted the welcome beginnings of domestic emancipation. In 1920, for the first time, the average domestic consumption of electricity exceeded 400 kwhr per year; this is clear evidence that uses other than lighting were being accepted. For a time the Company tried to keep some track of installations, but very soon abandoned this as hopeless.

The search for more significant items brought some surprising adventures to Edison. Application of electricity to the simpler jobs of heating in industry suggested a move into domestic cooking equipment, long time private domain of the gas industry. When manufacturers proved to be unresponsive to complaints about both price and capability, the Company itself, through a subsidiary, entered the manufacturing field with its own "cooking machine," to the dismay of the industry. This did not prove directly profitable to the Company, a fact to which the depression contributed, but it had the indirect result of shaking up the manufacturers with the demonstration of a market for an electric range. In the spreading suburbs, the electric company's lines went faster and further than did the pipes of the gas distributor, and in these extending areas, electricity became the undisputed leader in the modern kitchen.

The electric refrigerator needed no such direct stimulation by the Company. The manufacture of artificial ice was itself a relatively recent development, and the domestic ice box was the mark of affluence as late as 1910. It was a drippy mess at best, and the iceman, even when he came only every other day, still tracked up the kitchen floor, dripped water across the room and introduced a kind of crisis of domestic disorder.

In an early form, the electric refrigerator was a unit installed in the ice compartment of the old icebox, with a compressor in the basement. Increasingly, the economy and efficiency of the unit box became obvious. The first electric refrigerator on Edison lines is not dated, but in 1922 there were perhaps one thousand in use. The sale of this machine took little promotion.

As new conveniences became necessities, the simplest measure of the degree of their penetration through the economic levels was the kilowatt hours consumed. In 1921, the average domestic consumption was 408 kwhr per year. The similar figure, averaged over the greatly extended service area, and the greatly increased number of customers was, in 1947, just over four times as great per customer.

By 1947, electricity was a commonplace. The Company counted well over a half million (555,000) refrigerators in use in the area, most of them in the 771,000 homes which Edison served. The electric range, despite a troubled early start, numbered two hundred thousand. Obviously, electricity no longer meant just light to the householder; the Company used the phrase "electric living" to describe the new affluence.

Social historians have noted a twentieth century alteration in the position of women in society and in the home, an alteration in which the constitutional amendment providing votes for women was epoch making. One would be unwise, however, to limit the word emancipation to this context in matters political. Electrical appliances contributed enormously to this freedom and in a way this freedom gave substance to the hopes which women had expressed.

Sales, as Detroit Edison defined the function, was almost unique in the utility industry. The original service, as Thomas A. Edison conceived of his system, provided the customer with a "burner," as a burner was always available with the competitive gas service. But Detroit Edison, almost alone in the nation, continued to furnish electric light bulbs to its domestic customers without direct charge.

In the earliest days, when electric irons, toasters and fans came close to being the sum total of domestic conveniences, almost all were sold from the offices of Edison directly. Miss Sarah Sheridan, the charming lady who managed these offices, had at her disposal most of the persons skilled in electrical service within the city. As a good merchant should, she felt responsible for the continued performance of the goods she sold, and she invited customers to bring in these electrical gadgets for free repair as needed. No one ignored the fact that this assured prompt restoration of the equipment to the Edison lines. Even after these goods were common commodities in the retail trade, she and her successors maintained the broad custom of either free or "at cost of parts" repairs.

Around the city and the broadening urban service area, the attractive Detroit Edison offices were a symbol of personal attention and individual consideration to customers. In 1966, for example, there were seventy-two customer offices where a housewife could report interrupted service or faulty equipment to any of seventy-two branch offices, or by telephone to the central office, and be reasonably sure that an Edison service man would appear within the hour. In the early years, when electrical equipment was less reliable, the service call usually meant no more than a changing of a fuse. With more complicated equipment, including installed motors and heating units, and especially with many manufacturers providing that equipment, the task of the service man became more complex. Over the years, however, the Company had used prompt repair and maintenance service as a feature in promoting sales of special domestic items.

The exchange of electric light bulbs, the free or low cost repairs and, above all, the extraordinary service of the Edison man created a total package which continues to astonish newcomers to the Detroit Edison service area. Locally, customers refer to this as "free service," but this is an error Detroit Edison does not itself make.

Regularly, once or twice a decade, someone in or out of the Company will challenge these practices, and suggest that the Company could save large sums of money by curtailing or even eliminating these extraordinary services. An especially elaborate committee inquiry in 1959 estimated that the distribution of twelve million light bulbs cost the Company $1,750,000 per year. For many years, The Detroit Edison Company has been the world's largest private purchaser of electric light bulbs.[1]

The decision after every such reevaluation has been to keep the program. Edison can buy light bulbs for a third of what the customer would pay, and he seems willing to have that cost included in his light bill. Services are not "free" but so far they have never been challenged by either customers or a rate-making commission.

This reference to "free services" raises another interesting speculation. Costs of meter reading and collecting are not proportionate to the size of the bill, and the demand for these special services is certainly not proportionate to the household usage. The minimum bill in the current schedule of rates does not pay the costs of the minimum customer. It would be interesting to attempt to find what share of the total number of customers pay less than

A suburban office (1933) showing the "cooking machine" which the Detroit Edison developed and marketed. The other equipment was also of recent development.

Early electric washing machine.

their cost as realistically established. Beyond doubt, there is a subsidy for the small accounts which is buried in the general averages.

Especially in good times such as prevailed in most years since World War II, there has been a kind of rivalry between two departments of the Company. Production, preoccupied with lines and power plants, struggled to keep up with the power demands of the customers. Sales struggled to put on the lines more and more customers asking for more and more kilowatts. At times Sales came too close to completely overwhelming Production for Company comfort.

In 1946, Fred Compton was appointed general sales manager. Compton had worked in Purchasing since 1933 and was purchasing agent after 1940. This appointment was a deliberate expression of a desire for an aggressive and positive leadership in Sales.[2] Compton was thus one of the number of men who were brought forward in what can be called the Marshall-Parker period of leadership. In 1949, his title became vice president for Sales. In 1951, Edwin O. George was made commercial manager, and the next year he became manager of Sales. He was a ranking member of the department, and upon Compton's retirement became his successor.

To the people in Sales, characterizing the utility as "noncompetitive" was a poor description of the reality they faced. There are, it is true, some services which only electricity could perform, but they were few; there, the competition was not with others who could do the same things, but with vendors of other items who were after the same dollars.

Extension of electricity's role in the household would come from the introduction of new items, or from the substitution of electricity for other forms of energy. Such a substitution would involve in the future as it had in the past a competitive struggle in which price, convenience and reliability would all be factors. In such a competition, the customer must be convinced that the purchase of electrical equipment was to his advantage; in this sense, Sales must be quite conscious of the customer's point of view.[3]

It did not follow, however, that the interests of the Company were well served by every sale. In any given day through the year, the demand upon the Company will show wide swings in customer

THE DETROIT EDISON COMPANY
SYSTEM LOAD
PEAK DAY PER MONTH
MEGAWATTS
2500
2000
1500
1000
500
DEC '55
DEC.'54
JULY'55
JULY'54
12
2
4
6
8
10
NOON
2
4
6
8
10
12
AM
PM

electrical requirements; the peaks and valleys tend to establish a pattern which can be anticipated with reasonable accuracy. The Company would find it advantageous to promote those appliances which tended to fill the valleys in the load curve. Research in marketing was primarily directed to one or the other or both of these purposes. Load research was as important as was research in customer acceptance.

In 1950, the Sales department with the assistance of the Statistical division completed a year long study of the acceptance of equipment. The conclusion was that the electric range was still the critical purchase, and that the other equipment went to households which had first purchased the range. This was hopeful, since in the territory as a whole twenty-six percent of the homes had this convenience. Inside the areas where gas was the established cooking fuel, the percentage dropped to thirteen percent, and this indicated the region in which aggressive work might be profitable.

In March, 1950, Compton sent a telegram to 1,250 appliance dealers announcing the willingness of the Company to provide without charge a 240 volt line to any customer purchasing an electric range. This service had been given in promotion periods before the war, but had been suspended in and immediately after the emergency. Compton's hope was to increase the overall customer acceptance by twelve percent. The Company expected to make some of the sales from its own seventy-eight offices, but probably not over ten percent; all advertising stressed cooperation with retail outlets of the area.

This intensive promotion was applied to two other domestic conveniences. The electric clothes dryer was immediately given the highly successful campaign used for the range. The Company area had at the time only three customers in each one hundred using this convenience, and while this was twice the national average utilization, the Sales department resolved to increase it by ten percent in this five year program. Again the Company paid the costs of installation. The water heater was also added to the campaign.

The decision to actively promote the three items—the range, the clothes dryer and the water heater—in one campaign rested on the analysis of the load characteristics. Any one of these items alone, because its demand for electric power would last only for a small part of the day, would have an unfortunate effect if a large number were sold. The three items, added together in proper proportion, tended to fill the valley in the load curve. Sales determined to try to sell two ranges to each dryer and water heater over the five year period 1950–55.

This five year promotion was referred to in the Company as the load balancing program. In this period, the goal was to increase the saturation in electric ranges from 26.5 percent in 1950 to 37.5 percent in 1955; the goal was 18 percent for water heaters and dryers. In no case was the goal quite achieved, but customer acceptance of both ranges and dryers continued beyond the period, to achieve the figures by 1956. Unhappily, the water heater was more of a problem.

Some kinds of electrical equipment seemed to need little promotion. In the one year, 1950, the number of television sets increased from thirty-two thousand to one hundred twenty thousand, and by the end of two more years, sixty percent of Detroit's homes had television. Edison recognized this in the development of its own advertising program. The "Kitchen Carnival" originated as an audience show in the sales auditorium; it ran for several years from 1948.

It is difficult to find exactly when some of these additions to living comfort caught the fancy of the buying public. As late as 1949, the garbage disposal unit seems to have been something of a novelty, though soon thereafter whole areas were able happily to dispense with the garbage collector. The electric incinerator was not well accepted. The electric blanket, or its equivalent, was no doubt widely used, but statistics, both of ownership and use, are difficult to obtain.

Under the leadership of Fred Compton and his successor Ed George, the Company maintained an aggressive sales program. Sometimes it was in association with national campaigns by the industry or its allies, but more commonly its emphasis was on the local scene. George was a strong believer in the concentration of promotion effort. To this end he used television and newspaper advertisements, sought cooperation from retail merchants, employed highly commercial methods such as a truck mounted display which toured fairs and picnics, and even allowed door to door solicitation.

Aware of the value of observation and word of mouth advertising, Sales department leaders especially encouraged use of the equipment by Edison people in their own homes. Once Compton estimated, with some exaggeration, that the Edison employe household was one in one hundred in the sales area; if he could do it, he would make every one of them neighborhood demonstration centers. This desire led to an elaborate program of discount and special item sales which the Company made available to employes.

When Edwin O. George became president of the Company in 1964, the post as vice president for Sales went to another long time member of the department, Howard Stevenson. Their association was close, as had been George's and Compton's. Their immediate task was to interpret the recent experience of the department, including the five year load balancing program. With the conspicuous exception of water heaters, the goals were substantially met. The work of the department had been in this sense successful.

Sales found itself somewhat on the defensive before the other departments of the Company. The costs of the program had been high during the period of the promotion. Each range, dryer, and water heater required special installation, at company expense and added to the potential servicing costs for years to come. The question being raised was whether the business was worth the cost.

Even more critical was the question of load distribution. For example, three times a day at almost the same moment, the electric ranges would make their brief but quite significant demand for energy. Taken by itself the load distribution was not a desirable one. This the Sales department accepted, and even the five year program was shaped to this reality. The dryer, the water heater and the range, when the loads were combined, created quite a different total demand pattern.

These facts opened a whole new field of research in domestic utilization. By 1960, representatives of the Sales department were confidently saying that the problem of the domestic peak would be solved by the increase in the number of pieces of equipment which demanded power; the increased number guaranteed the desirable diversity in use.[4]

The logic of this argument opened wide possibilities, but there was immediate concentration on two problems which had priority. The water heater had been the part of the five year program which had not attained success, and it continued to be a problem, both in rates and in service. The other possibility which waited study was in space heating.

The major problem with the hot water heater was in charging for service. The steady demand and the rather large load seemed to justify a special rate, if it could be devised. Such a special rate would only be justified if the heater used surplus power at off peak or low demand hours. A simple timing device was developed which turned on the heating unit for ten hours at night, to provide a full tank of hot water in the morning. Unhappily, domestic habits did not always coincide with Company convenience, and customers complained of cold baths. Even the extension of the heating time to a possible twenty hours (1956) did not meet the complaint.

The arbitrary clock which controlled an on-and-off mechanism became increasingly unsatisfactory for the customer, even when supplemental heat (at a different rate) was made available. It also proved inadequate for the Company in meeting the problem of the new summer time peak. Air conditioning made its demands at unpredictable hours and for different lengths of time. The clock could no longer anticipate either peaks or valleys.

An ingenious solution was the development of a radio control. A cut-off was installed (1967–68) in the water heater in the customer's home which responded to one of ten radio tone signals given from Company installations scattered over the territory. The radio band frequency was assigned by the Federal Communications Commission.[5] The system was developed with the cooperation of Motorola, which provided both the transmitters and the switching units. The total cost was roughly ten million dollars. When installed, the radio control allowed cutting off 100,000 kw of demand instantly, and a machine to produce that much power on the peak would have cost much more.[6]

This power was sold at a low rate because it was interruptible, at the Company decision, for not more than four hours per day. In practice, unless extended, the interruption was for five minutes at a time. In most cases, the customer was unaware that his service had been interrupted. With this device, water heaters became part of that total load which could be shed instantly in emergency, in any given amount up to one hundred percent of this load. The system was in operation for many months, however, before the first emergency justified cutting off all the heaters at the same time.

The second problem, more baffling than even the water heater, followed the ambitious proposal that electric heat be substituted for all other sources of artificial heat in the residence. This involved competition with coal, fuel oil and gas, if the proposals could be made effective.

Electricity had long been a limited source of heat both in industry and in the home. Space heating, however, had been substantially confined to a kind of supplemental heating service in which efficiency was neither tested nor was it significant. A portable heater in the bathroom gives only negligible experience compared to the problems of space heating for a whole house.

As late as 1949, an executive in the Company responded to a question on this problem by saying that electric units could be put in warm air furnaces, but that electric heating in this climate was not advised because it would cost several times as much as gas or oil.

In other parts of the nation, because of climate or surplus low cost hydro-electricity, experiments were being carried on, and Edison could not afford to ignore them. Most promising at the moment was the "heat pump," most simply described as a refrigerating unit installed in reverse. In at least one form, the machine was reversible, to cool in the summer and heat in the winter. The first of these units was installed in Edison territory in 1946, and over the next decade something like fifty pumps were installed, in thirty-five houses and business establishments, and were the center of constant study and experiment by the Company.[7]

Rather quickly it became apparent that electric space heating, to be most effective, should be a part of the design of the building and not a mere substitution for another type of heat source. Although older type buildings could be remodelled and insulated, it was difficult and expensive. With properly prepared structures, it was possible to use the simple and widely acceptable standard base board panel, the glass wall panel or the heating cables in floor or ceiling. There were also experimental efforts at heat storage, as a part of load balancing, which were promising.[8]

The tremendous market now open for domestic (and commercial) electric heating challenged and excited the men in Sales. Conventional advertising was used but, for this kind of innovation, demonstration and word of mouth approval would be much more effective. As had been done previously, the department turned to the Company employes. A credit of three hundred fifty dollars was offered to any employe who would redesign his home for all electric heat, or would install it in a new home.

The most ambitious step in this campaign came just at the end of 1963, when a special rate was announced for the "all electric home." This rate applied if the basic heat and water heater were electric; if other electric services were used, they were also charged for at this base rate—a price which almost guaranteed that the home would be "all electric."

The all electric home would have been an unthinkable luxury only a generation earlier. But in the twenty years after 1947, a large number of them were built, and an even larger number of households had half or more of the electric conveniences. Electric heat was, in the late sixties, close to being cost-competitive with fuels and, for an increasing number of persons, the advantages more than offset the higher price.

The domestic market was important in the affairs of the Company. The annual report of 1953 noted that, among the companies serving the eighteen largest cities, only one had a larger proportionate domestic lead. The domestic sales continued to produce more than roughly one-third of the Company's revenue, and the trend seemed to indicate an increasing proportion of the revenue from that source. It was the stabilizing heart of the business. Not once since World War II, had the sales to the domestic customer business failed to show an increase, in spite of national economic changes and the erratic fluctuations of local industries in sales and employment.

The stability of this share of Edison's market justified the efforts to stimulate and extend it. The success in these efforts was a major factor in Edison's success. Between 1947 and 1968, the number of domestic customers was less than doubled; but domestic revenue multiplied by more than five times, and kilowatt hours sold were six times as many.

THE DETROIT EDISON COMPANY
ELECTRIC REVENUES BY CLASS OF SERVICE

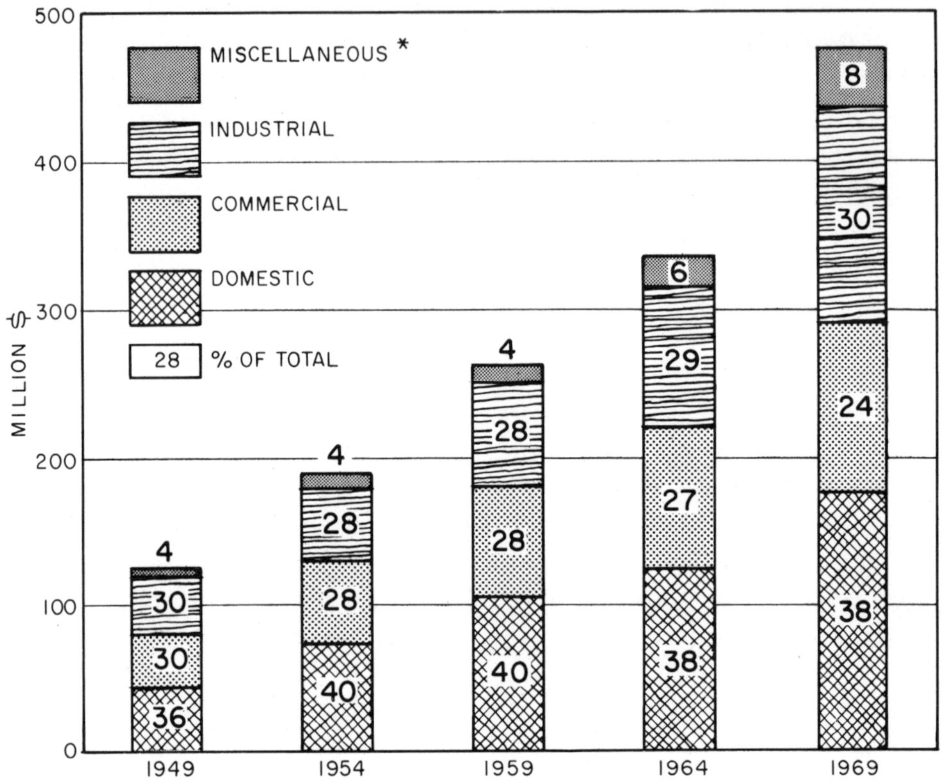

* ELECTRIC SALES TO MUNICIPALITIES, SALES FOR RESALE, ETC.

Sales was also involved with, and responsible for, the surging growth of industrial demand and for the large commercial sales, but the domestic customer remained the concern of most of the people in the department. The Edison customer could properly feel important.

The impact of this twenty year revolution is not dismissed in mere statistical tabulations, or measured only in terms of Company satisfactions. For the customer, and especially to the housewife, electricity in the home meant a further—almost an incredibly further—extension of domestic convenience.

The number of individual domestic customers was up, in round figures, from 740,000 in 1946, to 1,390,000 in 1969. Electric refrigerators were in practically every home, with the saturation well above 100 percent. The new ones had freezer space and were "frost free." In more than 625,000 of these Edison homes, the range was electric, with 49,000 new ones added in 1968 when the self cleaning oven was new. No one counted electric washing machines, automatic or old style, but the electric dryer spared the housewife one chore in 340,000 homes; in 1946 this aid was unknown.

The "comfort control" idea spread rapidly. The all-electric heated home was still a small minority even among those newly constructed, but in 1968, the number rose from 10,000 to 12,000. The centrally-cooled home was also still a small percentage, but at the end of these two decades, one-third of the buildings in southeastern Michigan were in whole or in part electrically cooled. So widespread was the use of air conditioning that the peak of the year's demand shifted from a December day to a warm midsummer day.

The count of radios had long since been lost. Television in 1968 was only two decades old, but already the number was beyond estimate. Multiple sets were common, and one-third of the customers had color television. And how many electric clocks, motors, controls, conveniences and necessities filled the modern house was only noted on those rare occasions when power was lost.

Perhaps the significant fact is, after all, statistical. The average of the domestic customers increased his annual use of electricity from 1,513 kwhr in 1946 to 5,697 in 1969; that is an increase of 3.75 times. In 1949, the Company had announced a price increase, followed in 1959, and following years by a series of price decreases. The considerable increase in the amount of electricity used by the individual customer gave him an increasing proportion at the lower range of the price schedule. Counting both forms of decrease, the average cost paid by the domestic customer per kilowatt hour declined more than twenty percent between 1951 and 1968, and in the economy of those years a price reduction was unusual news.

Farms and farm households were included in the category of domestic customers. Nowhere is the shift of the nation's economy more obvious than in this particular service. Only within recent times have federal statistics taken recognition of the divergence between the small farm and the large one, but the changing farm is reflected in Edison statistics. In 1950, the Company served 36,000 farm customers. In 1960, the number was 26,000.[9]

THE DETROIT EDISON COMPANY

DOMESTIC RATE VS CONSUMER PRICES
(1957-59=100)

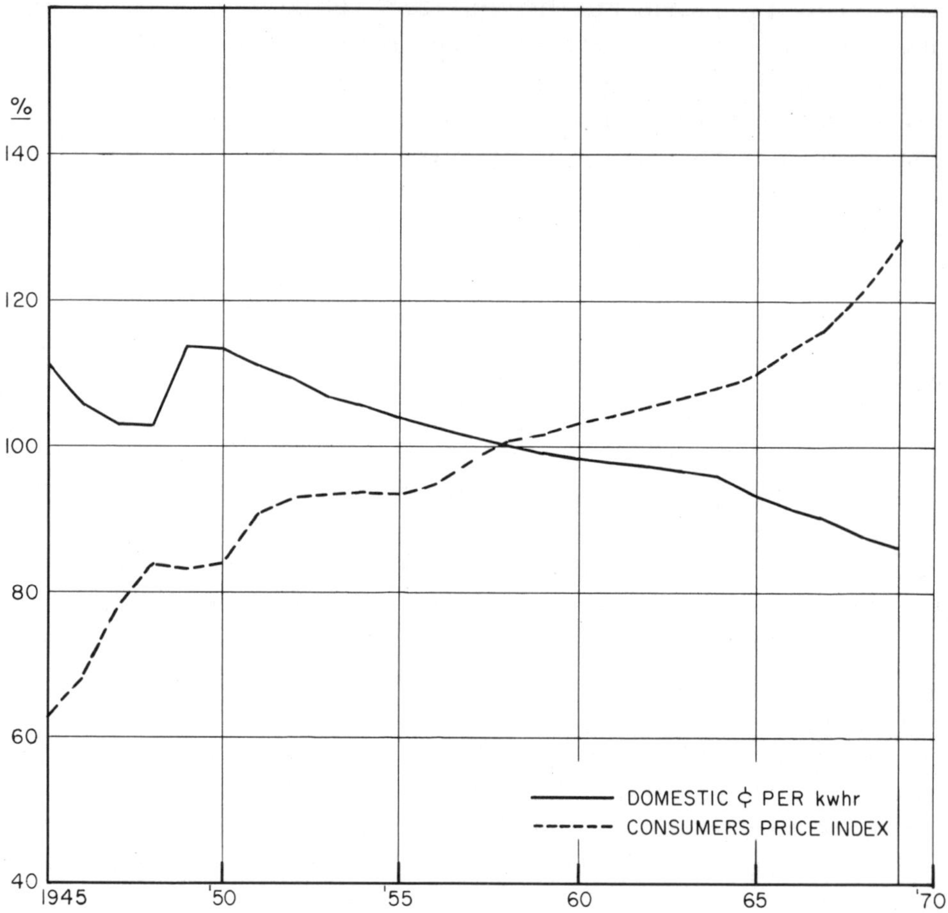

This is due in part, of course, to the elimination of the farms contiguous to the expanding urban area. A far more important fact is the elimination of the old subsistence and residence-farm, and the development of the modern quasi-industrial farm operation. The combination of small farms into the more efficient and productive larger farms introduced changes in methods of operation.[10] The Company estimated that its services equalled the labor of one man on each of these new and complex economic establish-

THE DETROIT EDISON COMPANY

SHAREHOLDER, CUSTOMER AND EMPLOYE GAINS (1945 = 100)

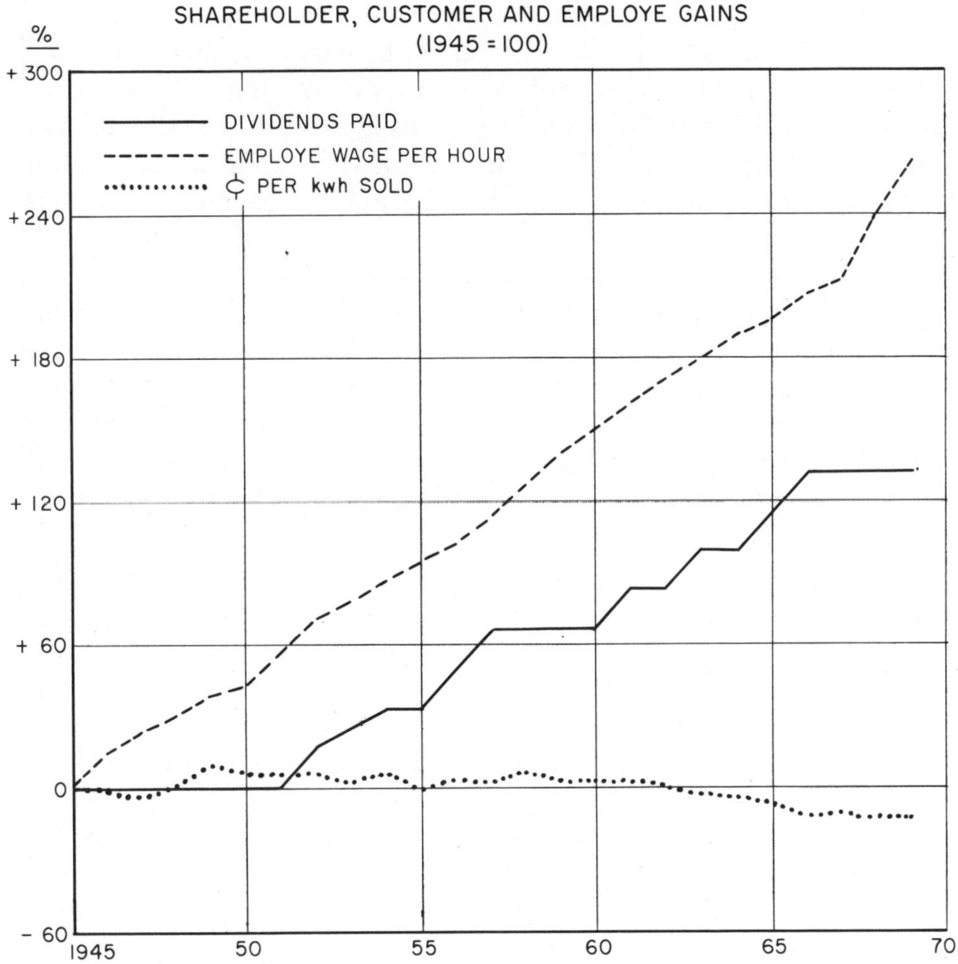

ments. It was nothing uncommon for a farm electric bill to run to sixty or eighty dollars monthly. The average farm customer used twice as much power as did the average domestic customer, and some farms' demand resembled that of a small industry.

Ideas developed primarily for the domestic market often proved to be applicable elsewhere, especially in the commercial market. The most rapid acceptance and development of electric heat came in connection with the construction of new school buildings, especially those of relatively small size. Here the advantage could be closely calculated. The elimination of the furnace room saved money and added usable space. Economy was especially noticeable when a school unit was being built, with unscheduled additions to follow; electric heat allowed such extension without either the over building or the rebuilding which would be required by a standard heating plant. The elimination of the furnace attendant on twenty-four hour duty could be a calculated saving, while constant temperature control could give immeasurable returns in health, comfort and cleanliness. The case was so persuasive, especially in rural and smaller communities, that by 1965 there were twelve schools of the all-electric pattern, with two hundred and twenty classrooms. In 1968, thirteen new all-electric schools were built, five more were under construction to be finished in 1969, and future construction plans would bring the total to sixty schools in this service area. The earliest of these all-electric schools had been in Utica, in 1958.

The explosive quality of Detroit's growth stimulated the building of supermarket type stores in the new suburban areas, and especially the development of the shopping center idea. At the time it was opened (1954), Northland was the largest such complex in the world, and promptly became an Edison customer with billings above three hundred thousand dollars per year. By 1957, Eastland had an almost similar demand, and The Detroit Edison Company served twenty-two other shopping centers. Twenty-four more were under contract or were in construction.

On first glance at the Company statistics, the number of commercial customers seemed to be increasing at a slow pace. In the city, and especially in the suburban areas, the huge shopping center had a tremendous vogue obviously at the expense of the small and independent shops and stores. Since many of these new centers found it economical to purchase power at primary rates for sub-distribution, the mere number of customers did not reflect total usage. In these new structures, the level of illumination both inside and outside was conspicuously high.

The extraordinary illumination which modern merchandising methods required naturally led to the association of light with heat. In some of the small offices of Edison, experimental demonstrations had shown the practicability of using light for heat. One kilowatt hour of electricity, used either for heat, light or power, is equal to 3413 Btu. Several sales offices, perhaps as many as sixteen, were heated by electricity, but the new installations at Oxford, Bad Axe and Howell used light, supplemented by heat only at extremely low temperatures.

It was only a step to the application of the idea to a larger demand unit, and with more precise calculation of all factors. A test installation in a single supermarket depended for heat on light plus the heat recovery from refrigerators and in-duct heating. This was a 1968 installation which looked promising. By the end of 1968, the "heat with light" was no longer experimental, and had been applied as an all-year concept to twenty-seven buildings.

CHAPTER TEN

SALES IN THE SERVICE OF INDUSTRY

Industry and the Reconversion / The Meaning of "Industrial" Power Sales / Detroit as a "Big Industry" Center / The Automotive Industry / Steel / Extension of Central Heating / The Sale of Industrial Steam / Total Energy Control / The Ford Contract.

THE HISTORICAL ASSOCIATION OF THE CITY OF DETROIT WITH THE automotive industry is one fixed in practice. Although the Board of Commerce was quick to argue that the city's industrial might was not confined to the single commodity. In considerable part, this contention stemmed from the feeling that an economy dependent on a single industry must be more vulnerable to wide cyclic movements. This was, of course, true when the industry was very new, and when the announcement of Ford's celebrated five dollar day was still a recent memory. The decade of the twenties improved the wages but did not reduce the long annual shutdown at the end of each season. Eight months continuing automotive production then marked a fairly good year. During the next decade of the depression that much work in the year would have been a Godsend.

Perhaps it was the fear of the return of days which led the city to emphasize so strongly the diversity of its industry. The industrial census, taken in 1947, was triumphantly quoted by a publication of the Detroit Board of Commerce (October 16, 1950) to prove this point. Detroit was the nation's largest center for manufacturing, for both automobiles and for non-automotive manufacturing industries. The number of workers, the wages paid, the value added in manufacture in Detroit's non-automotive industry exceeded the total manufacturing achievement of Cleveland, Boston, St. Louis or the entire state of Texas. According to the census, the value of automobiles was slightly exceeded by the sum of all the other manufacturing of the area.[1] Cautious observers would have noted that 1947 was too close to the war's distorted economy to be normal, and that shortages and disturbances both in material and in labor supply affected such figures.

Of greater significance for the Company, and probably also ultimately for the city's economy, were the records of Edison's own industrial sales. The largest annual production in the decade of the forties was in 1944, when shops and factories received for industrial purposes, and thus for war, roughly sixty-one percent of all power produced by the Company. In that year, the Company counted 561 industrial customers; in both the year before and the year which followed, industrial customers numbered 560.

The closing of the war in mid-1945 resulted in a sharp decline in sales, which carried forward well into 1946. Recovery in 1947 and 1948 brought industrial sales back almost to the level of wartime, and the year 1950 exceeded the wartime maximum by fifteen percent. The number of industrial customers fell from the war-stabilized 560 to 483 in 1948, and then promptly recovered to surpass the wartime figure in 1951.[2]

During the war itself, when plants on war production tended to run two or three shifts, the full use of machines gave a high load factor, mounting to 63.3 percent in 1944. Peace time production did not demand such full utilization of equipment, but, after a year or two, the load factor stabilized at about ten percent above the normal prewar figure.[3]

The industrial reconversion had not been painless, but it had been rapid, efficient, and relatively smooth. The city and its people shared in the resultant prosperity. Recovery in so short a time was reassuring, especially in view of the pessimism so often expressed during the war period.

Neither the Company or the city were prepared, however, for

what followed recovery. The wartime demand had taken the industrial sales above three billion kwhr in both 1943 and 1944. In the postwar recovery, that figure was again attained in 1948. The fourth billion sold was recorded in 1951, the fifth in 1955, the sixth in 1957. In round figures, industry purchased twice as much power from Detroit Edison in 1957 as it did in the recovery year of 1948 —and the 1957 figure was to be almost doubled in a following decade.

For the Company, this was exciting, and challenging. To the city and the industrial area involved, the phenomenon was significant and satisfying.

Despite the protests of those who somehow feel that bigness is wrong, and the efforts of the local business leaders to emphasize diversity, the fact must be recognized that Detroit was still a big industry town, and that much of the apparent diversity was in companies which were in reality satellites of the automobile industry.

A first glance at the Company's business accounts would not give that impression. Industrial accounts have never been more than a small fraction of the total number of customers; except for the very brief war period, the number stood from 1929 to 1948 at between four and five hundred. In the next twenty years, this number of customers doubled, and by the end of 1968 stood at 1,300.

This increase is more apparent than real. The rules of the Company required that each building which was separately metered be separately billed. A number of major companies, most conspicuously the motor companies, had several plants or buildings, and in the gross statistics became customers several times over. Chrysler, for example, was counted as a customer twenty times and General Motors twenty-three times.

The designation "industrial customer" is itself not precise, since it includes all those who find it to their advantage to purchase at the primary rate. Governmental agencies on various levels, utilities in gas and telephone, and the extensive office and retailing establishments found it advantageous to purchase at these industrial rates. The "industrial" customer purchased power at the distribution voltage of 4,800 or 13,200 volts, and owned and maintained the transformers and other equipment necessary to make the power usable.

For the larger user, the rate advantage was quite significant, though the exact comparison of the rates would be complicated. A general customer whose total electrical bill was in the level of four to five hundred dollars per month would be justified in inquiring about the effect and costs of a change to the primary industrial rate, and for a considerable number of markets and commercial establishments, the change proved to be desirable.

The number of customers on industrial rates, but not engaged in manufacturing industry, was thus rather large, and tended to grow both in number of establishments and in absolute size of demand per customer. In 1950, in round numbers the non-manufacturing share of the so-called industrial load was ten percent of the whole; in 1967 it had risen to fifteen percent. In kilowatt hours sold, this amounted to just a little over seven percent of the total sales of the Company.

The great bulk of Edison's industrial power flowed over its lines at distribution voltages to a limited number of customers in a limited number of industries. In 1952, twenty firms purchased 64.2 percent of the total manufacturing primary power. In 1957, it was 77 percent, in 1962 it was 78 percent, and in 1967, 80.1 percent. This is a surprising and increasing concentration. These twenty firms purchased 33 percent of all the power produced by Detroit Edison in 1967. The mutual dependence between the Company and the major customers is obvious.

A roll call of Detroit's major industries will produce no surprises. Among the customer groups, the automotive companies were first and steel was second. The three automotive companies and Great Lakes Steel each bought more than one billion kwhr per year. Wyandotte Chemical, largest of the chemical group which ranked third, purchased three quarters of a billion kwhr. In 1952, in the stabilized postwar world, eight large companies stood at the top of the list of customers; in 1967, the list was identical, except for one addition and one omission, and even the order was only slightly altered.

Beyond the big three automotive companies, the concentration in the areas of manufacturing was striking. There were three firms manufacturing steel, three in the chemical industry and three which manufactured and assembled steel products. There were two cement plants, two oil refineries, two paper manufacturing establishments, a brass works and a manufacturer of electrical goods. These twenty firms were locally dominant in their respective areas of production. All other companies combined demanded less than one-fourth the total energy demand of the twenty.

The Detroit Edison Company did not produce all of the electricity developed in the Detroit area, but its proportionate share appeared to be rising steadily. After World War II, it was estimated by Sales that the Company plants were producing roughly one-half of the area's total industrial electric power. This estimate for the immediate postwar situation is less reliable than the one fifteen years later, which set the Company's proportion of the total industrial electrical energy produced in the Detroit industrial area at two-thirds.

In many of these plants, the industrial process demanded steam, and of course, the requirements for heating could be large. With equipment already in, and with other continuing demands for steam, the operating costs chargeable to electric services were small; that is, the saving would be small if the electric production was discontinued. The moment of decision would arise when for reasons of obsolesence or inadequacy, the electrical equipment no longer served.

To a large number of these industrial plants, even those which supplied their own needs totally, the Company offered at most modest price what was called "stand-by service." The line, installed to the plant site by Detroit Edison, would be instantly available for use in case of accident or emergency.

The existence of the line was sometimes a decisive argument for its further use. Very often, the customer would find that all costs considered, the purchase of supplemental power by contract with Detroit Edison was preferable to any available alternative. On occasion, the entire power plant was closed down, and the existing boilers were thereafter used for industrial steam or for heating. In at least one case, involving Monsanto Chemical, the Company also contracted to sell steam in commercial quantities.

In any case, the calculation of the real costs, and the evaluation of available alternatives called for the skills of the sales engineer. In 1950, the reorganized Sales Department had twenty-one sales engineers whose job it was to reach the estimated 2,500 factories and shops in Edison's territory. These were already customers. It was the responsibility of the sales engineer to know the customer's business well enough to, hopefully, suggest alternative or additional uses for electricity at places in the industrial process which would be advantageous to the customer and at the same time profitable to the Company.

Except under special circumstances, the small isolated steam driven electric plants had substantially disappeared. Occasionally a proposal for a gas driven engine created some excitement, but a full comparison of costs usually proved purchased power more economical.

Sales accepted a second function also. Business organizations in Detroit were concerned with the stimulation of the industrial development of the area, a concern in which Edison shared. W. H. Hurley headed this activity in 1950 and found no less than eight agencies, commissions and boards all engaged in securing Detroit's economic future.[4] The Company itself followed up on inquiries from the outside about new businesses considering Detroit locations, and with some success.

The phrase "area development" at first meant nothing more than the response to requests for industrial information, often involving the problems of plant location.[5] In these matters, the Company did not restrict its efforts to the city. It followed with special concern those companies which, for reasons of space or inadequate facility, were preparing to leave Detroit; the Company was hopeful that the move would not take them beyond the Edison territory. In this interest, Edison joined the circle of smaller cities which were a part of Detroit metropolitan area, or were adjacent to it, and which were concerned with local economic growth. The most elaborate development in which the Company was directly involved was at St. Clair. Here the ability of the Company to supply both electricity and steam seemed to create a favorable situation for industrial use. The Company secured title to an extensive land area near its plant which it proposed to hold for a desirable development.

The automobile industry established a curiously erratic pattern during the fifties. In 1950, the nation's factories for the first time produced 8 million cars and trucks; the next year, the number was 6.8 million. In 1955 a new peak was established at 9.2 million. Throughout the decade, 1952, 1954, 1956, 1958 and 1961 were years in which actual decline in production occurred—with 1958 showing a sharp loss in sales.

The Detroit Edison Company's industrial sales showed the effect of this rhythmic pattern in automotive sales, but there is one significant difference. It is easy to note the effect of the automotive slowdown in 1951, but the decline in sales of electricity was not proportionate to the fall in car production; when automobile sales returned to the 1950 level in 1953, the industrial sales of the Company had risen from 3.7 to 4.9 billion kwhr.

This tendency persisted through and beyond the decade. The effort to maintain and increase production in manufacturing with a stable labor force led to technological improvement in the machines of production, and thus to increased use of electricity. For the nation as a whole, in the decade 1947–57, use of industrial power increased two and one-half times as fast as did industrial production of goods. In Detroit's complex and highly integrated industries, technical innovation, with almost certain increase in the use of electrical power, was the road to both efficiency and to new economies. Automation was only a step away.

In Detroit, Packard, Hudson and Graham-Paige remain only names in history. The automobile is a Big Three product. The largest in the industry, and one of the largest corporations in the world, is General Motors. The corporation was the product of the consolidation of a large number of firms, and it tended to emphasize individual firm identity. Subject to limited exceptions, the individual subsidiaries have large autonomy and are in their public posture competitive. Some policies are corporate wide, and one which most directly affects the Company is the policy of purchase of power when it is adequately available, as an alternative to the investment and hazard of private generation.

The plants at Flint were, of course, not in the Company territory; some of the power used in Pontiac was supplied by the Company through the local distribution company on contract. In the Detroit area, the elaborate headquarters' building, the vast G.M. research facility built later, and some twenty plants are in Edison territory. The combination makes General Motors the Company's largest single customer.

Chrysler was also the product of development plus acquisition. The largest single company acquired by Chrysler was the historic Dodge Brothers, in a purchase which to that day was the largest industrial purchase ever made. Chrysler was an extensive purchaser of energy, with the major exception being the old Dodge Main plant. This factory was built in the early twenties, when Edison was not prepared to offer adequate power in that then remote area. The result was a private power plant of considerable size which constituted for some years a continuing irritation to the industrial sales people in Edison.

The Ford Motor Company had quite a different history. Though it made some acquisitions, it was still at heart the corporation which Henry Ford had dominated so long. Until his death the company followed the policy of attempting absolute independence for the productive plant, with the autonomy emphasized by a high steel fence and a rigid security system.

This was a commitment to a system of administration which was already antiquated. The new regime which took power so promptly after the first Henry Ford's death began at once to decentralize operations, and to escape the limitations implied in the fence. As factories were created by Ford Motor in the general Detroit industrial area, they were usually beyond the economical range and capability of the River Rouge power system (which had been Henry Ford's personal pride, and his mark of industrial independence), and were, from the first, Company customers. The River Rouge plant, however, like the much smaller one at Dodge Main, represented committed capital, a depreciated but a still usable plant. The decision as to whether it would be modernized or abandoned was always one to which Edison's Sales department gave full attention.

The policy of decentralized automobile production which characterized the decade of the sixties, proved less critical to Edison than at first might have been expected. Even in production, the shift only slightly reduced the number of jobs, for the new plants elsewhere were increases in total production rather than subtraction from Detroit. The headquarters establishment remained in the Detroit area, as did scientific research and planning and other functions which tended to be much less responsive to the ebb and flow of automobile sales.

In 1967 and 1968, the percentage of the power produced which went to the automobile industry was higher than it had been in 1952, and in the sixties the boast was that thirty-three percent of

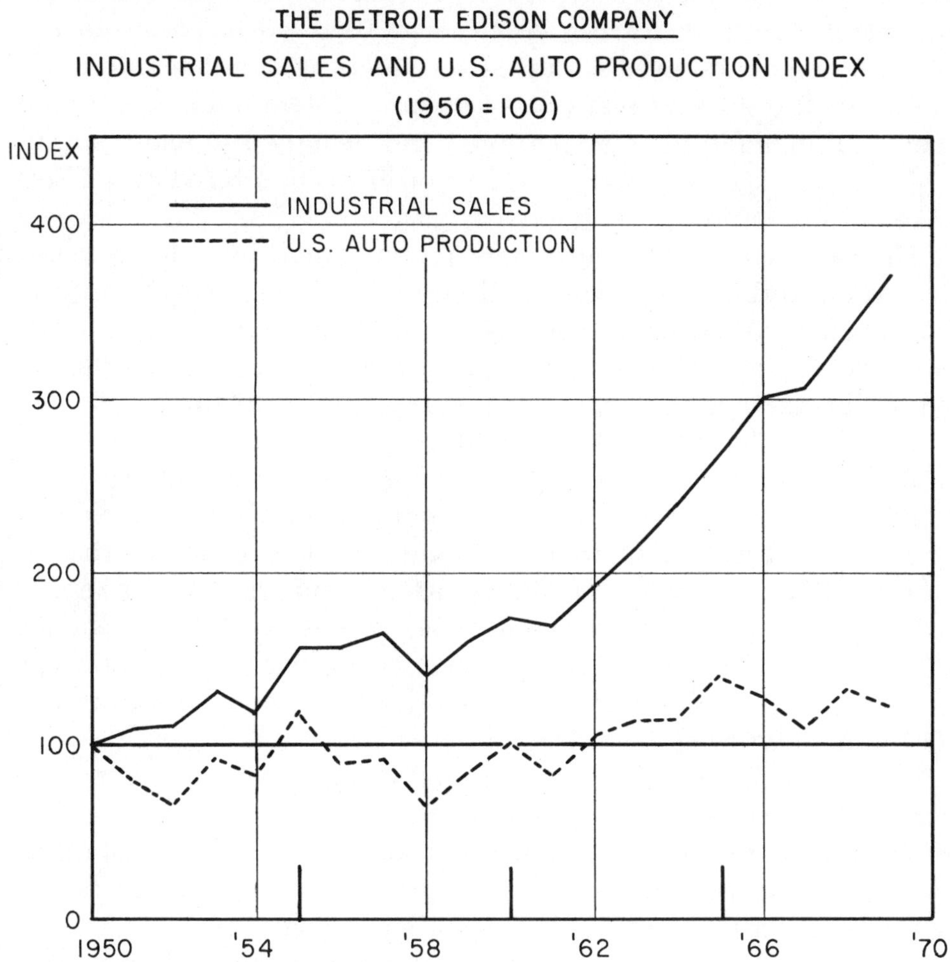

all American made cars were still being produced in the Detroit area.

The early history of the steel industry in the Detroit area is oddly brief. This is true, in spite of the fact that a hundred years ago, a present suburb of the city was the site of an innovation in the manufacture of steel contemporary with and almost identical with the famous Bessemer invention in Great Britain. No explanation is given in the records of why, before the Civil War, the operation was transferred to Cleveland.

This pre-Civil War venture was not followed up here, and in spite of the development in the area of extensive metal working facilities, the basic materials were produced elsewhere. The rapid rise of the automotive industry, and especially the industrial use of Detroit facilities during World War I served to point out the importance of the area's market for steel. As cars became more complex, the makers were in a position to demand steels which were tailor-made to specifications, and designed to exact standards. In this kind of work, the Detroit location offered an advantage which could be developed.

The decade of the twenties saw several small local firms begin operation and, by the time the depression descended, the efforts had been justified. In the somewhat confused early history it is possible to identify three companies or groups of companies which became so important in the area's later history.

The Great Lakes Steel Corporation was incorporated in 1929, and in November of that year became a subsidiary of National Steel. National Steel was itself a merger of a number of companies in Pennsylvania and Ohio and among others represented the interests of Hanna Iron. The Detroit area plant was in Ecorse.

McLouth Steel, incorporated 1934, represented some earlier efforts in iron and steel, and at least two other companies were subsequently acquired. McLouth operated plants in Gibraltar and Trenton, downriver from Detroit.

The third plant, smaller and more specialized, was called the Rotary Electric Steel Company until 1957, when it became a part of the Pittsburgh based firm of Jones and Laughlin. This plant in Warren was especially equipped to produce smaller quantities of stainless steel built to highly critical specifications.

Given the favorable local circumstance and the available market, the steel industry grew rapidly. World War II created an almost insatiable demand for steel, again with local firms as eager purchasers. These steel companies were thus prepared to accept the sharp local industrial increase in postwar demand, though growth also required an almost annual increase in facilities.[6]

Speed of installation, relatively low investment costs, and the ability to utilize new methods, all combined to persuade these companies to use electricity extensively. Great Lakes Steel was the only large steel plant in the nation which depended entirely on purchased power. McLouth, with the assistance and cooperation of the Company, pioneered in the use of electric melt and induction heating.

Edison Company officers pointed out in the annual report of 1965 that in the decade just closed Detroit's increase in steel had been at the rate of 60 percent as against a 12 percent for the nation, and that Detroit-produced steel was up from 5 percent to 7.5 percent of the nation's supply. The area leader in this industry was Great Lakes Steel, which was the Company's fourth largest customer, exceeded only slightly by the third of the Big Three.

The dream that somehow electricity could be developed as a costless by-product of some other activity died hard. The exhaust steam from the old reciprocating engines, and the unused surplus steam in the fired boilers, appeared to be a waste which led both New York and Detroit to ventures in central space heating. The newer turbine produced electricity substantially with no exhaust —certainly with none to use as heating—and well planned and managed boilers had no waste or excess steam. In Detroit, the creation of the first turbine plant at Delray with its condenser unit might well have ended heating service.

Very soon, however, heating acquired an important if subordinate place in the Edison scheme of things. The old so-called "isolated" plant was a steam boiler, an engine and a generator owned and operated to serve a given building. Such a plant was almost certainly inefficient and outmoded, and its existence constituted an open challenge to the Company and to Sales. But so long as the building or plant must maintain steam for heating purposes, with all the costs this implied, the electricity produced did become a kind of by-product, and the plants were hard to displace. The function of the Company's Central Heating department was to aid in the crusade against the isolated electric plant.

The Heating Division had never been an especially profitable enterprise; normally a dollar invested in its plant produced less income than one put in the electrical system. The competition with the privately owned heating systems which could be returned to service held prices well below the level theoretically and legally chargeable. Expansion in the heating service area occurred usually in connection with the expansion of electrical service. Twice at least, this kind of expansion had given to the Company an independent heating plant of such proportions as to imply continuation rather than abandonment of service.

Circumstances external to the Company dictated the growth in heating service. During World War I, when coal was so scarce, central heating was an unconditioned blessing; between 1914 and 1919, revenues rose two and one-half times. In the hectic prosperity of the twenties in Detroit, when skyscrapers first marked the downtown skyline, the dependence on Edison heat was such that many of the new buildings had no facilities for an alternative system. Revenues rose from one million dollars to almost $2.9 million between 1919–1929, with almost no increase in the number of customers. The largest number of customers ever served by the heating division (2,057) was in 1920. The depression years cut the number, but inducements in convenience and cost in wartime brought back the number in 1945 almost to its earlier peak. In the downtown area, probably 95 percent of the buildings were served by Central Heating.[7]

In organization, Central Heating had been a kind of autonomous unit in the Company. In 1950, selling steam was added to the other functions of the Sales department. In the postwar years, the steam customers were served from one of four interconnected boiler plants. The Willis plant was located not far from the location of old Edison Illuminating Company Station B, where heating service had begun. Two small and troublesome plants in the crowded downtown area had been superseded by the much larger modern Beacon Street plant. The plant on Congress stood near the site of the purchased Murphy Heating Company (1914) while the Boulevard plant, purchased directly from General Motors (1928) had been created to serve the General Motors building. It was in uptown Detroit and was thus three miles north of any other heating service.

The interconnection between these four major plants created a central heating system which was second in size only to New York's. The Company was able to meet the demand from other buildings created in the New Center area, even when it came from major buildings comparable in size to the General Motors building.

A half mile to the west of this New Center area, the new multistory Clinic building roughly doubled the size of the Henry Ford Hospital. Faced with the need to rebuild its heating plant or make other arrangements for heat and power, the administration accepted the offer of the Company to provide both heat and electric power. A one and one-half foot pipe under city streets provided the service (1952) to what was potentially the largest single customer.

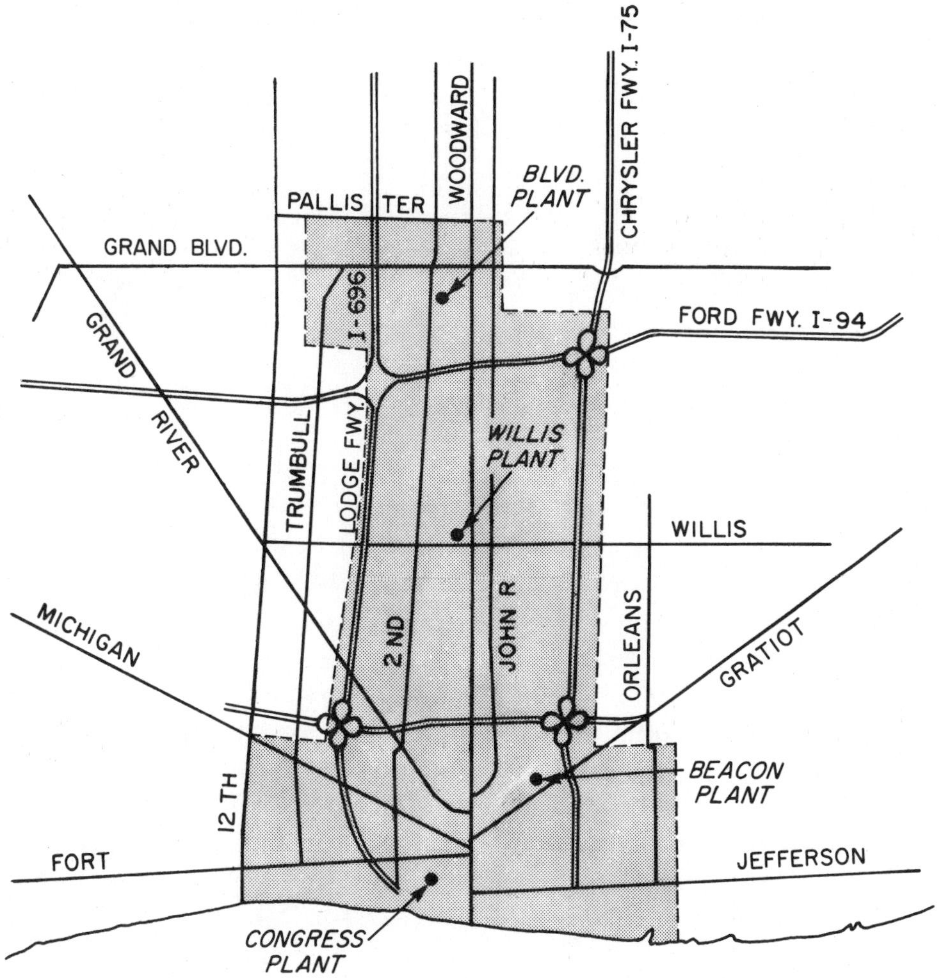

CENTRAL HEAT AREA

The story was repeated several times as other customers developed along or near the line. The Burroughs Corporation closed down and then wrecked its heating power plant. Wayne State University found the service highly advantageous for buildings on its campus and closed down the furnaces in its Old Main building. One after another the center area hospitals abandoned their now expensive private plants and when plans matured for a great Medical Center-Medical Campus in the neighborhood of the hospitals the huge complex turned to Edison for both heat and power.

In the older downtown area, the story was much the same. The Stroh Brewery, east across the projected Chrysler expressway from the heating plant, was reached in 1957 by a tunnel 1,386 feet long, the first such construction since 1928. It was intended to serve also the new housing development in that area as it developed.[8]

With 95 percent of all downtown buildings, and with all major buildings connected with the service, area growth imposed a continuing demand. In 1959, it was necessary to add the first new boiler in a heating plant in eleven years, this one at Beacon; it added 25 percent to the total capacity of Central Heating, which by that year had 43 miles of mains and 1,900 customers.[9]

Over the years, in the war for the elimination of the isolated plant, the aggressive sales division had found significant assistance in the service of Central Heating. Unhappily, in the area to which industry had moved, that service was not available. The continued existence of the isolated plant related as much to the problem of steam as to electricity.

There were, in 1963, a total of 845 industrial customers who, combined, purchased half of the total power produced by the Company. But Sales noted unhappily the existence of a huge potential market beyond possible reach. The estimate for the year was that 1.8 billion kwhr would be produced in Edison area by such private plants as Ford, U. S. Rubber, River Raisin Paper and the like.[10] These power plants also provided steam for industrial or heating purpose, and Edison could offer them no alternative.

In the neighborhood of the old Delray plant, a number of major plants or other places of business were clustered. They could not be served from the regular lines of the Heating Division. Delray was in an unfortunate position in 1955; its old and less efficient turbine-generators were put on a stand-by basis, but the low-pressure boilers associated with them were still usable. The Sales department proposed to sell steam to Delray's neighbors, and made contracts with Timken-Detroit Axle, Revere Copper, American Brass, Detroit Harbor Terminals and old Fort Wayne. It was a new area for this kind of operation and, indeed, for men in Central Heating who had charge of the operations when installation was complete. The revenues promised to be a half million dollars per year, with capital costs small.

The experience at Delray, where the sale of surplus steam had brought important supplemental benefits, seemed quite worth examining. This provided a way in which the limited scope of the Central Heating service area could be expanded, and made every power house potentially a source of steam as well as of electricity.

This was not a new idea, but it did extend operations markedly. In the heart of Detroit's heavy industry area, customers who looked to Edison for electric power found it convenient also to become customers for steam. Great Lakes Steel was supplied from Delray and River Rouge, while Monsanto Chemical had long depended upon Trenton Channel power plants. These contracts were directly with the production departments. The Morton Salt Company, in the Marysville area, was supplied with steam from that plant on contract.[11] The arrangement for the commercial sale of steam from other plants awaited only the solution of the problems of cost and availability.

A dramatic and potentially significant breakthrough came in 1963, when the Company announced that it was taking over the management and operations of the power plant associated with the Pennsalt Chemical Company.[12] This plant had seven turbine generators, all of them small, with a combined capability of 37,000 kw. The oldest, dating from 1926, was worthless and only one was less than fifteen years old at the time of the announcement. For Pennsalt to expand its general operation, as it had planned to do, the antiquated plant would have to be replaced. This capital cost that company would be glad to avoid.

The contract with Edison called for the supply of the total energy to the chemical company. The existing plant would be maintained with fuel and operatives all supplied directly from Edison. This plant would provide steam to the chemical company. The decision on which and how many of the generators to use would be determined by Edison on the basis of comparative costs. For the moment, about one-half of the capacity could be in use. The major part of the load, and certainly any increase in it, would be carried from the Company's plants and large machines.

A larger and more complicated operation came the next year, when the Company took over the two power plants of the Wyandotte Chemical Company. Here there was a name plate capability of 57,000 kw, but again machines were old, small and inefficient;

there were thirteen turbines in the two plants.[13]

The contract indicated the reasons that impelled Wyandotte to make new arrangements; the expected load would be 1.7 million pounds of steam per hour and 80,000 kw demand. To provide these capacities, the Company agreed to install two new boilers, and the electrical facility which would allow the import of Edison electric power. The potential demand was beyond the capacity of the existing plant.

The third such contract, with the Port Huron Paper Company, was announced in July, 1966. At the time it was stated that the expected demand would be double the capacity of the plant as it stood.

In each of these cases the qualified employes in these plants had the option of transferring to Edison or accepting other jobs with their older employer. Of the thirty-four men operating the Pennsalt plant before the takeover, twenty-three chose to accept Edison employment. The three total energy contracts involved power, steam, compressed air, water, and presumably any other form of energy demanded. The contracts for total energy negotiated with Pennsalt, with Wyandotte Chemical and with Port Huron Paper seemed to be an innovation which might have wide application. For over a decade, and in some cases for a longer period, Detroit Edison had worked with these companies, developing methods by which their own production of energy could be supplemented by Edison, to the advantage of both parties.

Total energy contracts seemed to be a method with wide applicability. In November, 1965, a list was carefully prepared of sixteen customers who generated all or part of their electrical supply.[14] Several were under investigation, with the total energy conception as a possibility. One on that list, Port Huron Paper, was shortly signed to such a contract. With United States Rubber, the negotiations were extensive, but produced a standard contract only.

Despite the early optimism, the total energy concept seemed to have limited application, and sometimes brought unexpected problems. If the Company could have retired these ancient machines at once, calculated savings would have been secured. The pressure for maximum electric production, however, forced continued use of plants and machines which were no more efficient because Edison now owned and operated them. These industrial plants, because of the man power required and the attention demanded by staff, were a constant concern to management.

The Ford situation was more complicated, if for no other reason than that their power plant at River Rouge was larger. The contract for mutual assistance between Ford and the Company, an arrangement which long predated World War II, provided for a 25,000 kw interconnection. In 1952, this was superseded with a 75,000 kw interconnection. In 1954, Edison agreed to sell 110 million kwhr to Ford, at Highland Park, Dearborn Engineering and Administration, and to purchase an identical amount from the Ford Rouge Plant. The two companies were thus selling surplus power—Ford in winter and Edison in summer—to a customer whose needs then fitted that schedule. The system of exchange, with payment for excess either way, has continued to work with modification. Substantially the relations with Ford resemble those with another power company more than with the usual industrial customer.

CHAPTER ELEVEN

A BACKWARD GLANCE

The Changing Scene / Hydro-Power and D. C. / Some Items From History / Pontiac and the Other Cities / Suburban Street Lighting / The Warren Service Center / The End of the Decades.

BY 1965, TWO DECADES HAD PASSED SINCE THE WAR. SINGLY, THE YEARS may have seemed uneventful, but the twenty, combined, had brought profound changes both in the city and in the Company.

Downtown Detroit was changing. There had been an earlier boom period, chiefly in the decade of the twenties, when the face of the city was altered. The depression of the thirties ended this construction—sometimes with the building still in process—and the long period of war which followed confined construction to other areas and purposes. The prosperity of the fifties, extending into the next decade, produced a renewed activity in building and for a second time the Detroit skyline was marked with the skeletons of skyscrapers under construction.

As usual, progress cost the city some treasured landmarks. The first tall building of downtown, the old Hammond building across from City Hall, gave way to a bank. Detroit's first real steel structured skyscraper, the Majestic building, on the opposite side of the square, was scheduled for destruction in 1962. The ancient Detroit

Opera House and the equally old City Hall, which had faced each other across Campus Martius for most of a century, also fell to the demands for progress.

Inevitably, the question was raised about the ancient building on Washington Boulevard which since 1886 had been the visible symbol of Detroit Edison. Amid so much which was new and modern, it was almost defiantly old. This was the building which had housed the first Edison machines in Detroit, and here both Alex Dow and Henry Ford had worked before the days of their fame.

Inside the Company, there has been criticism of the building's retention, as a bit of unjustified sentimentality. The building has been reproduced at Henry Ford's Greenfield Village, machines and equipment included, and there were those who argued that the museum was the proper place for it. The argument had even been advanced that it misrepresented the Company's highly modern technology. There have been offers for the land, however, and they have been considered. It could be that the building survived only because the offers were thus far inadequate.[1]

Change had become the accustomed way of life. For the industry and the city, these decades of achievement could be reduced to statistics and charts. To the men to whom the Company had been a career, these figures had meaning, but so also did some events which hardly merited public notice.

The ancient Delray plant, the first construction of the corporate Detroit Edison, had had a brief life. Its funny little spinning-top generators, its troublesome boilers, and above all, the inadequate foundation which let the stack develop a kind of drunken tilt—all these marked an early and somewhat unsatisfactory effort.

Delray #2 (1908–12) was a different thing. It remained for many years a production center for the Company, and even after the building of Conners Creek and Trenton Channel, Delray was for the men who worked in the power plants a kind of sentimental symbol of technological excellence. But in 1957, the veteran building was demolished. The plant had long since been stripped of its generating equipment, and the scrap steel would bring $40 per ton. There was also a saving in taxes.

No one could object to the decision, and no one did. But to the hundreds of men who had trained and worked there, this demolition of the first great power plant marked the passing of the first generation in the modern company, and that was important.

Years before, when the Company was young and for the first time was reaching beyond the confines of Wayne County, there had been an exciting development in Huron Valley. The Company had purchased land and flow rights there, and had installed a series of five dams. Water power was the economic justification but Alex Dow took pride in the beauty of the small lakes, and insisted on proper development and landscaping.

There were five small plants, at five dams, plus ponds owned by Ford and a paper company. They stretched some thirty-five miles through the Huron Valley, and covered about 2,100 acres. They had been economically important when created, but by 1959 the 2,200 kw capacity was a microscopic part of Edison's production.

The ponds and the hydro-plants associated with them dated from the first decade of the Company's existence, though the last plant, the one at French Landing, was delayed to the decade after World War I. The enthusiastic first planning was the Company's reflection of the wild enthusiasm for hydro power which characterized the high day of Progressivism and the Roosevelt brand of conservation.

But hydro-power proved to be too expensive, even at no cost. The total contribution from these plants did not justify the attention they required, and certainly would not permit modernization. The decision was made that they be decommissioned, with the last going out of service in 1959.[2] The Valley and the rights owned by the Company went up for sale, valued at roughly one million dollars.

However, there proved to be a limited number of persons purchasing ponds. The ultimate sale of Barton, Argo, Geddes and Superior to the city of Ann Arbor (1963) was at a price roughly half the first estimate of worth.[3] The city of Belleville was even less able to deal with purchasing Edison Lake and the dam at French Landing and it remained a troublesome and unproductive asset.[4]

Changes in technology naturally changed Company practices. When Edison Illuminating was created, it used, and viewed as its greatest asset, the famous Edison patents and the associated system. The technological basis assumed the use of the old three-wire 110–220 volt direct current system. This system in Detroit, as in every city which boasted an Edison Illuminating Company, became the standard for the inner city.

When the alternating current proved so adaptable, especially for distribution over longer distances, the expanding areas used the new a.c. system, and in many cities left the electric service in two unequal parts. In the central area, the Company (1930) began a system of duplicate distribution, to allow the customer either service. New customers tended to accept the alternating current.

At long last in 1962, the only remaining direct current substation was eliminated, and after seventy-five years d.c. distribution was eliminated. Detroit Edison no longer used any of the patent rights for which it had once paid so dearly.[5]

There were some people then who argued that the name Edison should have been dropped, with the abandonment of this most characteristic feature. Some also felt Detroit could be dropped from the Company name, for its territory had become much larger than the city. In both cases, the historical association with the name seemed more important and surely was more valuable than the change dictated by this hard logic, and nothing was done.

There were a number of excursions into fields which had once seemed to be related to electricity. These ventures usually came to the Company as a part of a more normal expansion of electrical service.

When the Company purchased electrical service in Port Huron in 1919, the Company acquired the local gas distribution service also, and somewhat reluctantly maintained the business. In 1950, when natural gas became a commercial possibility, the gas service was sold.

In the Thumb territory, the Company acquired (1936) a rather strange assortment of other businesses along with the electrical systems. The central heating service in the small towns of Lapeer and Bad Axe had been associated with the village power plant when electricity was locally produced. Under modern circumstances, the service could not be justified, and in 1950 both plants were closed.

The situation at Vassar was a little more complex, for here not only heating but the village water system was involved. The heating plant was not essential and was closed down in 1950, but water service, although to fewer than one thousand customers, was vital. The Vassar water system survived—an odd fragment in a huge corporate holding, and a kind of a test, as it was viewed, of the capacity of the Company to perform on so small and personal a scale.[6]

Many business connections ended just because they were no longer needed. The new type diesel locomotive on the railroad made the ancient electric "mules" which had moved both freight and passengers through the railroad tunnel between Detroit and Canada obsolete. With their passing a contract for tunnel power which had stood since 1914 ended. It was not a big contract, but it was all that was left of Alex Dow's dream of an economic invasion of the Canadian area. When this was blocked, the tunnel contract lost its unique importance.[7]

Some problems seemed to solve themselves, given time. Even the rural cooperative, in the Thumb area, so small and yet so irritating to the Company leaders when it was launched, became accepted in time. When its original plant proved to be inadequate to the growing load, the REA board decided in 1955 to purchase stand-by service from the Company instead of a new machine. The experience proved satisfactory at 500 kva, and by 1965 the board in Washington approved the local contract by which this cooperative bought eighty-five percent of its load from the Detroit Edison's lines. Although the lines of the two systems continued to cross and duplicate with small regard to reason or economy. Substantially, the REA was the purchaser of bulk power for redistribution over its own lines.[8]

Expansion had carried the Company as far as Pontiac by 1908, and from that date Pontiac remained a continuing problem. Despite Edison's efforts, the small local plant was sold to a predecessor of Consumers Power, and the two companies seemed doomed to economic warfare. Similar situations elsewhere had developed into conflicts of critical dimensions.

Only after several years was an agreement reached; probably both parties then saw it as a temporary one, but it survives, providing for a curious salient into the general area of Detroit Edison service. Edison offered its services outside the 1910 boundaries of the city of Pontiac, and offered to sell electricity to the holder of the franchise inside old Pontiac for distribution by the local company. At the time, the demand in the restricted area was so small as to make uneconomic any other method of supply. But through the years since, Pontiac has been a frequent point of friction, as the two companies were forced to adjust to changing local conditions, and to changes in technology. Consumers' continuing purchase is listed by Edison with other customer sales for resale.

There were several other so-called wholesale customers inside the Detroit Edison territory. The city of Croswell owned a distribution system, but purchased all its power. Three cities, Wyandotte, Sebewaing and Clinton, long committed to their municipal systems, maintain plants. In each of these cases, the city produced or purchased power both for its own use and for resale to citizen customers. The decision to become an Edison purchaser was usually made when increasing load forced a consideration of alternative methods.

Detroit itself was served by the Public Lighting Commission's plant; a competent staff and an efficiently operated plant had provided for much of the city's electrical service, including practically all of the street lighting. This plant had been good, but it suffered as did many of the smaller private power plants when compared to the efficiency of the Company's large producers.

In 1962, the Board of Water Commissioners retired the two old steam powered machines at Springwells, and this ended on-site steam pumping in the city. Edison shared the responsibility with PLC for future electric power for pumps. In the planning of the sixty mile, ninety inch water main from Lake Huron to the city, Edison of course assumed the pump load.

For many years, the Public Lighting Commission had provided a major part of the power used by the municipally owned street cars, with Edison providing power in areas where PLC could not do so economically. The gradual phasing out of the electric street cars, and then of the electric trolley system allowed the equipment of the city-owned plant to be used to meet the enlarging demand for lighting service.

The PLC's central problem was its Mistersky plant, and it illustrates the dilemma into which publicly-owned enterprises can sometimes fall. The plant's ancient equipment demanded a decision about the future role which the plant would be expected to play.[9] In 1964, the Company and designated city officials began a joint study of this plant and its economic potential.

The Mistersky plant had two relatively modern machines, rated at 44 and 50 mw, and four original machines, all dating from the twenties and rated at 20 mw each. The Company too had machines, in Marysville and Trenton, which were of the same age, which it planned to retire or reduce to emergency use only. Presumably, the PLC should also plan to do this.

The issue at Mistersky was whether these four machines should be replaced with a modern machine big enough to meet the Commission's present and prospective demand. The study report, dated March 22, 1966, and signed on behalf of the study group by H. C. Reasoner for the Company and H. W. Pearson, General Superintendent, PLC, offered alternatives, but recommended the kind of arrangement which had become a familiar one in similar industrial situations. The old and expensive machines would be retained, but on a reserve basis; the newer machines would operate at their most economical and efficient level, substantially around the clock. Edison would buy any surplus thus produced, and would supply any amount of power demanded above the capacity of the operating machines. This joint operation of the PLC system with the Edison system would provide economies for both. But the city did not accept the proposal.

A second issue arose when the Mistersky plant was involved as a violator of the city's own rules on air pollution, and when the Commission found itself facing the necessity of spending unavailable capital to make the required modifications on antiquated equipment. In the spring of 1968, Mayor Jerome P. Cavanagh was once again informed of the Company's continuing willingness to cooperate in any solution deemed feasible.[10] But a decision was apparently difficult politically and was postponed.

Public lighting was Detroit Edison's responsibility in the rapidly growing suburban areas. There had developed among these communities, a sharp competition in public service and civic improvement, and an urban type, modern system of street lighting was a most visible evidence of achievement. Edison shared this pride and, in fact, also shared the cost. Although there was no policy decision to this effect, the contract price for these lights probably never provided proper return to the Company.

In 1960, the *Readers Digest* public service award for the planning, development and installation of street and highway lighting went to fifteen cities and towns in the United States, and Detroit Edison served six of them. They included Highland Park, Hamtramck, Ecorse, River Rouge, and both Grosse Pointe Park and City. The next year Grosse Pointe Woods entered into a seven year contract to achieve equal excellence.

In so rapid an expansion, the Company's main worry was its ability to perform. Demand was often beyond its normal capability. The unavailability of qualified men to plan and install these lights, either as Company employes or with private companies on contract, created an embarrassing lag which became almost habitual between the receipt of the order and the complicated installation.

There was a public safety factor in this press for increased street lighting, and especially after the disorders of the summer of 1967 the demand mounted. In that year, the Company served one hundred thousand street lamps, and the tendency was to increase the size and effectiveness of such lights.

In the early twenties, the new General Office building on Second Boulevard was constructed, and shortly after its completion a wing was added. Just before World War II, the Service Building was erected on Third Avenue directly opposite the General Office building.[11]

The old brick building on Washington Boulevard had been Company headquarters for three and a half decades. With these two new buildings available, the office space in the Washington Boulevard building reverted to some specialized and collateral uses, or even housed related civic and public organizations. The sales office on the first floor remained a familiar Company symbol in downtown Detroit, and provided an access to the Company for all business purposes.

By 1950, however, the space available for administrative and supervisory personnel was again inadequate. There was some discussion of transferring the total complex out of the inner city, but both economic and civic arguments were used against the idea. There was space for an additional building on the property already owned, and adjacent to Grand River. This was currently a parking lot behind the General Office building. With the city's cooperation, the area was enlarged slightly by shifting First Street's location as it entered Grand River, and plans were begun for an impressive sixteen story office building.[12]

These plans were completed by 1952, but any start of construction was made impossible by the Korean War. When building again became feasible, construction costs seemed to make the project unsound and it was again delayed. A management committee made an examination of space utilization, and temporary accommodations were found for some offices in other buildings owned by the Company or rented for the purpose.[13]

Any plan to reopen the proposal for this office building downtown was ended by the decision to purchase what came in Company terminology to be the Warren Service Center. The Ford Motor Company had transferred production of the Lincoln car to newer facilities, and the old Lincoln plant at Livernois and Warren became available.

In November, 1955, Edison directors authorized an offer of $4.5 million for the fifty-two acre property, which included seven major buildings. The entire complex was not available at once, for Ford continued to occupy some space, and some had been leased to other organizations. The property also required extensive alterations and modernization prior to complete occupancy.[14]

The office space at the Warren Service Center was larger than the two downtown buildings combined, and the parking space was notably more adequate. This acquisition allowed the transfer of whole units from crowded conditions, and encouraged more efficient organization for handling supplies, material, and especially heavy equipment.

According to a preliminary estimate, activities from twenty-six locations could be transferred to the new service center, and perhaps twenty-one of these properties could be sold. By 1960, the move to the new center was sufficiently advanced to allow this action. On January 24, 1961, C. F. Ogden requested permission to offer for sale thirteen pieces of property with a value estimated at $2.5 million; although not all of this was released by the transfer to the new service center.[15]

The postponement, and then the ultimate decision not to erect the building on Grand River, left this small but attractive corner vacant. During the delay in plans, the area had been planted with grass and attractive shrubs. It was a beauty spot in a rather bleak part of the city. On motion of Robert R. McMath, director, the Company engaged Marshall Fredericks, a well-known sculptor, to plan a monument to Alex Dow for this location.[16] This monument was dedicated, with simple ceremony, in June, 1959.

The leadership group which Walker Cisler had built in the days following World War II had proved itself competent and dedicated. The two postwar decades had been marked by solid achievement. There had been occasions which had momentarily seemed critical, but, happily, most of these difficulties were a result of the pressures of growth, and management proved capable.

These two decades had not been marked by spectacular innovation, either in the field of production or in business organization. The ebb and flow of economic activity had been within acceptable limits, and after the somewhat hectic first postwar years, had settled to a pace of comfortable and predictable annual increase. Even the recession following 1958 had been within the limits of anticipated variation.

The sharp increase in demand in 1962 could be read as a recovery movement, and perhaps also the equal growth in demand in 1963, but when it was increased in 1964, and again in 1965, it was clear that the pace of the industry had altered. The new situation would become a crisis if this growth pattern continued even a few years, and if Company reaction to it was not prompt and effective.

ALTERNATIVES AND DECISIONS

Introduction

The historian is accustomed to completions, and the story he tells usually will run to accepted conclusions. In this sense, the story told in Part Three, because it is unfinished, is hardly proper historical material. We do not here have the privilege of being wise in hindsight.

Let this be the excuse, if one be needed, for a departure from the rather simple historical narrative. It is possible to identify specific problems or clusters of problems. Each of these has emerged naturally within the Company's history, and each has commanded attention to an important degree in the decade of the sixties. Because they are current problems, we can know with what knowledge and purpose the management moved, and what was the action taken. The outcome we cannot yet know.

The function of the first chapter of this section is to state, or perhaps restate, the fundamental rules or conditions which must shape decision, and then to apply those restrictions to the situation at hand. The great urban region centering about Detroit will make massive demands on the Company for energy in its most flexible form. Inside the Company, some person or persons must make the critical decisions of where, when, how much, and at what cost? Such decisions, once made, can seldom be reversed, and thus the responsibility is an awesome one.

The chapters which follow discuss, as separately as their interknit character will allow, problems with which the Company must deal. Specifically, we propose to discuss (1) the emergence of an electrical complex in which Detroit Edison is but a contributing part, (2) the commitment to the increase in the production of energy, (3) the concern with fuels and combustion.

We propose, in other words, to write the "history of the Company to 1975." The decisions made after 1962 commit the Company to that distant date and beyond. They define now its capability to meet the obligations which will develop in the decade ahead. By the nature of things, this is also the future history of Greater Detroit.

CHAPTER TWELVE

THE IMPERATIVES, AND THE DECISION MAKING PROCESS 1964 - 1975

The Parameters of the Industry / The Economic Climate 1963–67 / The Administrative Structure / The Board and the Decision Making Process / Directives from Outside / Statistics and the Forecasts.

THE FUNCTION OF MANAGEMENT, IN THE FAMILIAR LANGUAGE OF THE economics textbook, is to combine labor, capital and resource in production so as to achieve the greatest potential economy and efficiency. It is a generalization which, however much abused in practice or distorted in experience, still describes the aspiration of the competent man of business.

If the problems were as simple as the textbook illustrations, then indeed the tasks would be easy. In reality they often call for the gift of prophecy. Labor's cost and productivity, the price and terms of capital accumulation, and the condition under which resources—including transportation—are made available, these are only within broad limits subject to management's control, or even its prediction.

In the electrical industry, there exist a collection of limitations which are more or less distinctive and which tend to restrict the area of managerial decision, even beyond the usual. These boundaries condition the response which the Company can or must make to the new, and it will be obvious in specific descriptions that these limitations are not new to the decade.

The central purpose of an electrical utility is to provide energy to its customers. Qualities of reliability, availability and price are the performance standards by which a customer will measure a given company, or by which a company will judge itself.

Unfortunately, these qualities sometimes, even often, stand in contradiction. The expenditure of large amounts of money to make available large reserve power, and thus to change reasonable to absolute reliability, would result in prices beyond the acceptable. The practical problem is to attain both reliability and availability at a price which is tolerable, accepting as a penalty something less than perfection in each separate quality.

The equation is not a simple one. Electricity cannot (in significant amounts) be stored. The customers demand can be met only if equipment is both sufficient and available on the instant. Since that demand comes with little or no warning, there must be at all times an amount of capacity in reserve sufficient to meet any unanticipated load.

No one knows what the demand on the Edison system would be if every light and every machine on the lines came into use at the same instant. Years ago, men tried to keep such a record of all connections, but it was abandoned as impossible and in any event unimportant. The best informed guess is that at no time is the actual demand more than one-eighth (twelve percent) of this connected load.

This being so, it is clear that there will be wide swings in the usage, and that the curve of demand of the day or the week or the year will show deep valleys and high peaks. It became possible to recognize these swings and to anticipate them with reasonable accuracy. Year after year, on some December evening, when holiday lighting of stores and enlarged domestic utilization coincided with the last of the day's industrial demand, the Company hit the peak. During the war, daylight saving distorted this pattern, and the peak temporarily appeared when factory work began before daylight. It reappeared as a summer peak in the sixties, with extensive use of air conditioning equipment.

This peak defined the Company's obligation, for if electricity cannot be stored neither can it be improvised. For three, four, or even five years, the planning, financing and building of the entire system had been aimed at this one fifteen minute period in a late December afternoon, or in a summer heat wave.

Obviously, providing plant and equipment for brief peak utilization was very expensive. On one occasion, Cisler estimated that the top fifteen percent of the productive capacity was actually used only 348 hours—eight percent of the year. The industry uses the words "load factor" for the figure which indicates in percentage terms how close the Company is in average utilization to its peak demand. Even in the heights of war activity in 1943 and 1944 it was only 63.3 percent.

Second only to the imperative requirement that power demands be met instantly and fully, is the equally insistent need for consistent utilization of capacity. Power not sold can neither be saved nor returned, and constitutes an obvious loss. Some of this loss is of course inevitable, but management must try to hold capacity to a minimum, consistent with reliability.

This then was the dilemma before management. Inadequate investment in capacity meant that the peak load would not be met, with embarrassing and even dangerous results both to the Company and to the dependent industries. On the other hand, investment beyond effective utilization meant a waste of capital, excessive costs and different but also dangerous results.

The third imperative is imposed by technological change and the consequent factor of obsolescence. In a literal sense, a turbine-generator never wears out. With proper maintenance and replacement, the machines could continue to run indefinitely without substantial alteration of efficiency and without increased costs. The trouble is that this is not good enough. Machines become technologically antiquated. New machines may make such savings in fuel and labor costs, per kilowatt hour, that economy requires the discarding of the still functioning old machine. Modernization is a requirement; management decision must establish the rate and the extent.

The distance between laboratory demonstration and commercial application is still a long one, but the technological possibilities must now include more than mere microscopic improvements in efficiency in established methods. In the past, the press toward larger size boilers, to higher steam pressures and temperature, and to more complex machines sought, fundamentally, refinements of tested methods. The current generation of utility men had witnessed the breakthroughs to totally new fuel sources, and must anticipate as possible even revolutionary methods of production.

The unknown which reduced all these factors from a calculation to a kind of intuitive judgment was time itself. It was not enough to know that demand will grow. The critical estimate was the one which said when and where in the future a given demand would be present.

With the increasing size of both machines and plants, and the complexity of transmission, the time lag between the decision to build and the availability of the facility tended to increase. It took from five to seven years to plan and create a power plant; the industrial customer could install a demanding process, or build a new plant in much less time. Success required that management, well in advance, anticipate a maximum demand, calculate with precision when it will develop, and build to satisfy it with certainty, but without wasteful overbuilding.

These requirements imposed on management can be simply stated, but meeting them was not a simple exercise in arithmetic. The records of the Company, both recent and long range, were searched for clues and analogies, but historical parallels are never complete, and offered no assurance that events would be repeated.

The immediate evidence was easily at hand. The sharp economic slowdown of 1958 had an effect which extended beyond the year. There was a moderate increase of general business in 1962. The year 1963 was even more prosperous. These years, however, proved to be the turning point for the economy of the nation, the city and the Company.

The recent rumor that industry was moving away from the Detroit area seemed worth examining, and sales began to report an unusual statistical series. The figures told of industrial movement into and out of the Edison territory, of accounts closed and opened, and of jobs lost or gained. The statistics for this period of returning prosperity showed that the migration of both industries and jobs was most emphatically to, rather than from, the area. Further evidence of prosperity was the sales department's report on delinquency in customer accounts, which had declined in gratifying degree.

In a speech before the New York Municipal Forum in April, 1964, Cisler emphasized the recovery from the national economic recession and from the period of doubt about Detroit's future. The automobile industry was spending, he said, a half billion dollars in capital improvement in the area, and other business expansions under way were even more impressive.[1]

Ford, in 1963, had added the first new assembly plant in the area in a half decade. In 1964, General Motors announced expansion in six different plants in the Detroit area. The automotive production broke the long-standing record of 1955, did it again by a larger margin in 1965, and, with the single exception of the year 1967, annually exceeded the figure of the preceding year.

The impact of such prosperity was marked in new production records for the Company. In 1965, on June 28, the record for the all time peak was broken for the first time in the summer. There was a second peak in September, but neither record survived the winter. Prior to September, 1965, the Company had never sold as much as 70 million kwhr in a single day; in 1966, that figure was exceeded 215 times.[2]

The demand for electric power began to press upon supply and, if one projected this upward trend only a little way, the capacity of the Company to serve at peak points would be tested. By 1965, every available machine, regardless of age or efficiency, was on the line in the effort to match capability to demand.

The more critical problem before management, however, was not the immediate need, but the meaning of this kind of growth for the future of the Company. The acceptance of a projection which showed a constant increase of this order would have been itself difficult, but it appeared it was also necessary to accept a curve which implied that a constant acceleration had become the new norm, which reduced it all to an act of faith. Projection of such a curve gave a forecast which, in spite of all statistics, seemed to defy reason.

Yet, this is precisely the position in which the decision makers were placed. The assumptions about the lasting power of prosperity were questioned, by implication, in the annual report for 1963, and almost annually thereafter the report of mounting sales and constantly increasing peak and production contained a caveat. The assumption that future conditions would duplicate the past and stimulate growth to an equal degree was frequently chal-

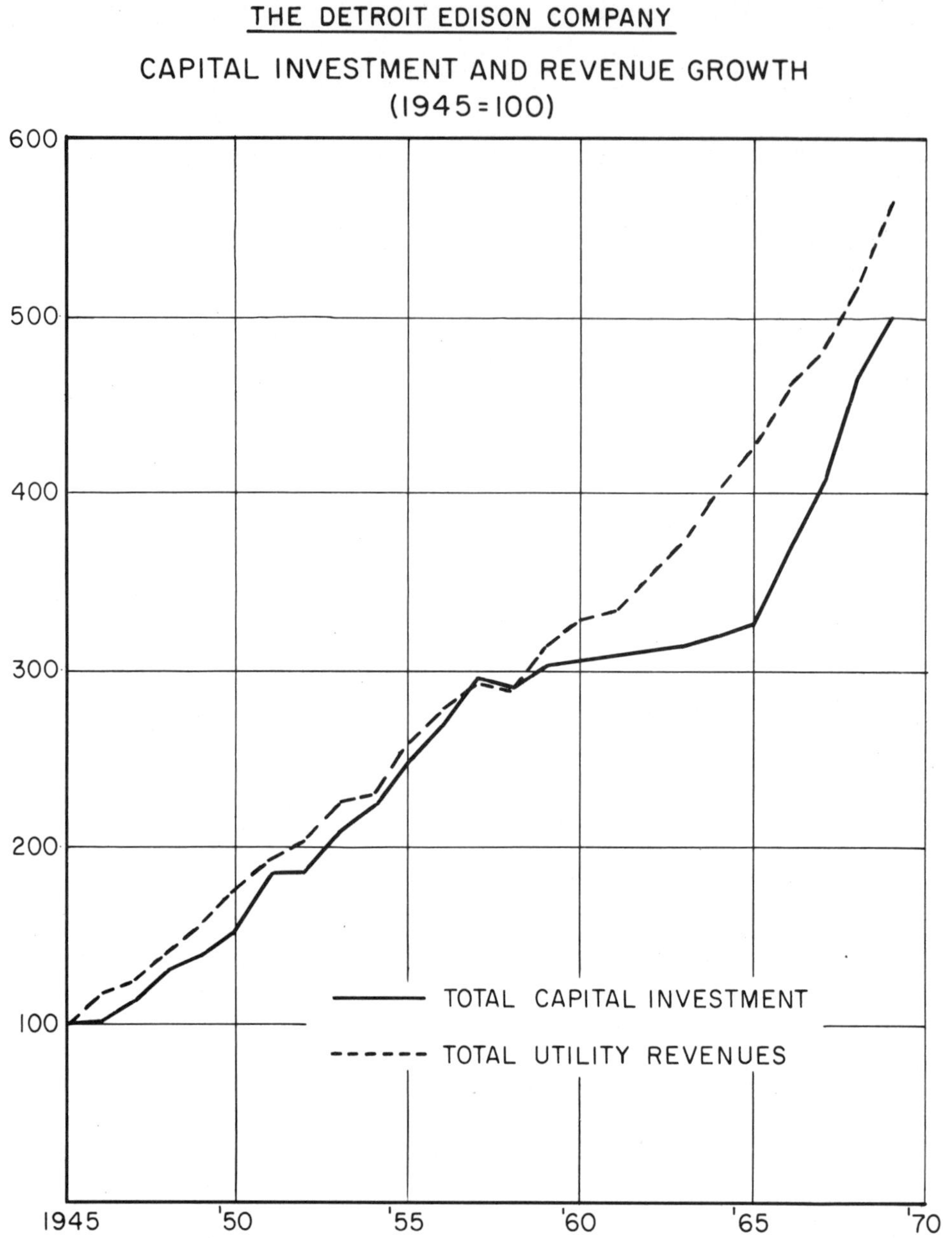

lenged, only to have these doubts overwhelmed in the evidence of continued increase. Despite the obvious contradictions, this was the foundation on which managerial decisions must be built.

The decision making power was vested in two groups of men, who, because their responsibilities and restrictions differed, presented differing sets of criteria. One group was the professional corps of the Company itself, the men in line responsibility or on staff duty whose collective judgment would be transmitted to

Cisler or George to be modified or more commonly authorized for action. The second group was the board, with the legal authority but without the immediate contact with the problems and possibilities.

In 1964 the officers and those who worked with them in line and staff responsibility were an experienced group. Substantially they had been in charge for a decade. When Walker Cisler was made president in 1954, Landrigan was executive vice president. A series of personal adjustments and retirements allowed the election of eight vice presidents in the years 1954–57. The team thus remained intact to 1964, with the exception of Hurn's and Landrigan's retirements.[3]

In 1964, the board of directors elected Cisler chairman of the board, and thus gave him both the title and the authority which Brown had held.[4] This action may have precipitated a number of changes in the organization. The most conspicuous of these was the promotion of Don Kigar and Chester Ogden. Kigar had been executive vice president since 1961, and his promotion to the office of president continued him as second in command. Ogden, who had headed the purchasing department, also with the title of vice president, became administrative vice president with extensive responsibilities. These appointments clearly were intended to mark out a new pattern of administrative leadership.

George Porter was senior among the vice presidents in tenure, for he had been in charge of engineering and production since 1952, when Paul Thompson retired. Eldred H. Scott had worked with Landrigan, had been his successor as controller, and had been vice president since 1955. Edwin O. George had for a long time directed the sales department though recently the designation had been vice president in charge of marketing.

These three men were appointed senior vice presidents on April 19, 1965. The title proved to be one of convenience, however, and was not used again. Indeed, the title of executive vice president was reinstituted, and as of April, 1967, both George and Porter were given that title.

In the early years of Cisler's presidency, the executive vice presidency had been chiefly honorary, and had been used to signal the ending of a distinguished career in the Company. This may have been the intent for Porter, for he had been ill, and he retired only a few months after he received this designation. But for Ed

George, the appointment had quite another outcome. In midsummer, Don Kigar resigned from the presidency of the Company. Ed George, who was several years away from retirement age, was named to succeed him in that office.[5] He and Cisler now were the senior officers of the Company as well as the ranking officers.

The form of the new organization which was being developed showed itself first in the departments affected by these promotions. Both as senior vice president and as executive vice president, George had retained responsibility for the department he had led so long. As president, that of course was no longer possible. Howard R. Stevenson, who had worked with George in marketing for many years as assistant vice president and vice president for marketing, was now elected executive vice president for marketing. In 1969, when his retirement was announced, the title became executive vice president. Thomas J. Peterson, who had been assistant vice president in the department, also retired in 1969.

The responsibility for leadership in the whole marketing division went to George Lahodny, who had been made vice president in 1967. He had earlier served, both before and after joining the sales department, as a personal assistant to Walker Cisler, and this assignment he continued in the new position as head of marketing. Paul A. Duker was then moved up to vice president.

The naming of George Porter as senior vice president allowed the appointment of Harvey A. Wagner as vice president; Wagner had been assistant vice president and manager of system development, within the production department. When Porter retired in 1967, Wagner succeeded him as executive vice president for production, and then as executive vice president. William G. Meese succeeded Wagner as vice president in 1967, and as executive vice president for production in 1969. Since Wagner had announced his retirement for 1970, Meese's designation as his successor was important. In a matter of months Meese was elected president on the retirement of Edwin O. George.

There was also some modification in organization, and some clarification of responsibility within this department. In 1969, George J. Brown was assistant vice president and manager of construction, and J. W. Drummond was assistant vice president and manager of operations. In 1970, both men were elected to the position of vice president. At the same time, Harold C. Reasoner became executive vice president for production.

Ogden's shift from purchasing to administration was followed by the election of Marshall Pease, a veteran member of that department, as vice president for purchasing. His successor upon his retirement in 1969 was Robert W. Lundgren. The vacancy created by the resignation of Sylvester F. Leahy in 1968 was filled by the appointment of Robert E. Schwab as vice president and manager of employe relations.

Substantially, the Detroit Edison Company had created a new supporting team behind Cisler. The transfers, the retirements and the promotions which sometimes seemed almost transparently contrived had had the intended result. The able team, created early in the period of Cisler's chairmanship, had been replaced after a decade or more by another young and able team, from Company ranks.[6]

The major departments, administration, marketing, employe relations, and purchasing were in the hands of a newer generation. The average age was less than fifty-five, and this renewal of the administration after 1965 was strikingly similar to the one which had occurred soon after 1953.

Among the Company officers, the transfer of responsibility to younger men was even more obvious. Harris R. Symes, secretary, Robert W. Hartwell, controller, and Arnold J. Benes, auditor, were all named in 1968, and Phillip D. French, treasurer, and Robert O. Wagner, general accountant, were named the next year. The election of William G. Meese as president occurred in 1970. The average age of the six officers was just over fifty-two.

The creation of the new administrative group raised again, and for the time effectively settled, the question of internal promotion as against recruitment. The characteristic career of the group had been totally within the Company, and the path to the top had usually been marked by changes of assignments and highly varied experiences.

The conspicuous exception was in the election (4–15–68) of James P. McCormick, the only man within the Company's modern history to be named vice president without any Company experience. In fact, his appointment tends rather to confirm the rule. He came to the Company from a career in education. A professor of English at Wayne State University and recently vice president for student affairs, McCormick had served also on behalf of the university in the joint program with Edison for the development of the Doxiadis study (see Chapter Seventeen). Within the Company, his responsibilities include relations with educational institutions outside, some participation in such programs within the Company, and in general the representation of the Company as it develops experience and responsibility in an area not familiar to such companies.

In a few other cases, when special skills or unusual experience was required, the Company has been willing to recruit from the outside, but this is infrequent. In Cisler's opinion, it would stand as something of a reflection on the Company and its leadership if it did not consciously attempt to discover and develop the skills it needed.

The board of directors, as the legal ruling body and the representatives of the shareholders, must either confirm all decisions, or delegate that decision making power, with such restrictions and limitations as it chooses to impose, to the officers and staff.

The board is reelected each year at the annual meeting of shareholders in New York. In fact, the renomination and election of a director is routine, and board membership usually continues until death or resignation. The result is *pro forma* election, and a surprising record in membership tenure. It sometimes happens, in such a system, that time and circumstance will terminate several board members' tenure careers at near the same time, and this had happened here.

By curious chance, the members who had to deal with the critical problems of the mid sixties were in large part new to the board and even to the utility industry. The transition had been both sudden and drastic. The first indication came of an alteration in the board's membership in 1955. Fink and Crispell offered their resignations at the November meeting of the board. The next year, Burr and Russell W. Symes resigned. Parker's sudden death in 1957 was followed by the death of Charles T. Fisher, Jr., in 1958. In 1960, the long and eventful career of James F. Fogarty came to a close with his death. The next year, Lawrence P. Fisher died after a brief term as director. In 1961, William G. Foster (elected 1956) resigned, and in January of the next year McMath died.

In 1956, the membership of the board was increased from eleven to thirteen.[7] These two new appointments, plus those required by death and resignation, meant that between November, 1955 and

July, 1958, eight new members joined the five who continued from the older board. In 1962, the board size was again increased, this time to fourteen; five new elections in 1961 and 1962 left Brown, Cisler, and Tompkins as the only members whose tenure on the board ran back to the time of Cisler's election as president in 1954.

Tompkins remained a member of the board, a veteran who was much respected, and whose advice, especially on financial matters, was constantly sought. It was obvious, however, that the other directors who had originally represented New York, were not replaced from New York as they retired. Membership on the board went to Raymond T. Perring, Detroit Bank and Trust, and to Lawrence P. Fisher. Following the death of Fisher, a younger member of the family, Charles T. Fisher, III, a senior officer of the National Bank of Detroit, succeeded him. The Bankers Trust Company of New York continued to serve Edison in New York, but the shift of directors seemed to some people to be a kind of assertion of Company autonomy which, even if unconscious, was not less important.

The most conspicuous grouping on the board would be the men who, as customers and citizens, represented the local community. One of the early appointees of this group was Harlan Hatcher of Ann Arbor, well-known as president of the University of Michigan. Richard C. Andreae was president of the Yale Woolen Mills of Port Huron. H. Glenn Bixby was president of the Ex-Cell-O Corporation. George M. Holly, Jr., was a director of the Holly Carburetor Company. Though his association was with local retailing rather than with industry, J. L. Hudson should be included in this group. The new appointments from within the Detroit business community were in no sense provincial. Each of the businessmen and industrialists on the board came from a company which, while Detroit or Michigan based, was of national importance.

Walker Cisler had been drawn more and more into the current of national and international affairs. He accepted responsibility in professional and civic organizations most generously, and in so doing, broadened his acquaintance with the nation's business leaders. From this larger circle, he could recommend election of individuals who could serve the Company well.

Frederick H. Mueller had formerly been in business in Michigan, as head of a Grand Rapids industry, but when elected, had been serving as Secretary of Commerce under Eisenhower. Daniel

F. Gerber was a highly successful baby foods manufacturer; his major plant was in Michigan, but Fremont was not in Detroit Edison territory. William C. Foster was an officer and director of Olin Mathieson; he severed his connection with that concern and resigned his directorship in Edison when he returned to public service in 1961. Gerber continued to serve until his resignation in 1966.

Detroit Edison's heavy involvement with the effort to develop its Fermi atomic energy plant will explain the election of two men, and also their interest in serving. Robert F. Bacher, provost, California Institute of Technology, was a physicist with special experience in the creation of the atom bomb and with scientific developments thereafter; he had been a member of the U. S. Atomic Energy Commission, and held extensive advisory responsibilities in Washington. General Kenneth D. Nichols retired from the army after an extensive scientific career. He had managed the wartime Manhattan project and, after retirement from the army, was general manager of the U. S. Atomic Energy Commission for two years. His first association with Edison was as consultant in planning the development of Fermi.

The Detroit Edison Company had become, as a result of size and importance, a national rather than a provincial institution. The new board was national in scope, with widely diverse knowledge; the expert opinion on so many matters which was available within the board was a valued asset.

The very breadth and scope of the board reduced its capacity to deal with the day to day operations and with the particular. There were no utility men on the board, except the chairman, until 1964 when Don Kigar joined Walker Cisler, followed by Ed George in 1967. The tendency was for the board to accept the judgment and recommendations of staff and management on operations and other local matters. The talents and competencies of the board thus tended to supplement rather than duplicate the specialized skills of the staff and officers.

The decision makers within the Company have found their function on occasion preempted or in some degree restricted by a directive or a regulation from outside. Regulatory agencies, on one or another governmental level, have been concerned with corporate financing, intercompany relationships, marketing practices, labor practices, minority employment, plant safety, public security safety, aesthetics in highway and zoned areas, plant location, and fuels. Sometimes the results of unrelated actions were odd. But rarely did conflicting jurisdictions and purposes come so sharply to view as at the Monroe plant, where one agency approved stacks to 800 feet, to dispurse flue gas, while another limited the stacks to 500 feet, in the interests of aviation safety.

The transfer of so much of the decision making from Company responsibility to public agency or commission has intensified the Company concern with the process, as well as the result.[8] Agencies or commissions, even when part of the executive branch of government, conform to practices of the judicial process, with hearings, witnesses and elaborate testimony. Because any challenge of the results must be to a court, the tendency is to conduct these in a way which the courts would recognize as due process.

Counsel found themselves involved in conferences concerned with the day to day operation of the Company. Increasingly, the issue of what should be done came to involve the question of what could be done. The discussion of a policy or even of an action involved both the question of its wisdom, given the situation, and of its relation to law and the regulatory agencies. There was no alternative to this, for the legalistic regulatory atmosphere required the lawyer's skills and methods.

Harvey Fischer of the Detroit law firm of Fischer, Sprague, Franklin and Ford had long represented the Company as its Michigan counsel, and he had worked with the firm of Sullivan & Cromwell, New York counsel, in proceedings in Washington. Over the years, the interests of the Company had taken more and more of Fischer's time. He, as well as the current New York attorney, sat as legal advisor with the Board of Directors. He met with the management council and was there an articulate and forceful participant in discussion on the staff level. He or his firm represented the Company in matters of regulation before public authorities and the courts, and he shared in the preparatory conferences on policy and method. Fischer was the personal friend of the chairman of the board, and enjoyed his full confidence.

Decisions were being made on three levels, by the Board of Directors, by the officers and the staff, and by known and unknown public authorities. The reconciliation and coordination of them all was the special responsibility of Walker Cisler as the chief executive officer of the Company, and of the president who after 1964 was also a member of the board. Sitting with the chairman as advisor, Harvey Fischer was in a position to participate with these top officers in all levels of decision.

Men in the industry have long used as a rough rule the generalization that production doubles in each decade. In the Company, a superficial examination of the figures from sales would confirm the rule. The postwar years of 1946–47–48 were doubled in 1956–57–58, and these again were doubled in the corresponding years of the next decade.[9]

The examination of the growth figures in detail complicates the matter. If the industry doubles in ten years then the annual growth, cumulated, is at 7.2 percent. In the five years up to and including 1955, the average annual increase was 9.3 percent. There was a period of relative stability, 1956–1961 inclusive, in which the annual increase was only 3.5 percent but from 1962 through 1968 (ignoring as abnormal the riot year 1967) the increase was 9.5 percent. On the basis of this kind of experience, the forecast of a doubled output each ten years seemed both modest and inexact.

Since the days when Cisler as chief engineer had shared in its creation, the load forecast for the next year and for the immediate future had played an important role in the Company's plans. In 1963, Edward R. Moore retired from the Company; his place as chairman of the forecast committee was filled by the appointment of Harold Reasoner. This was an appointment which recognized his contributions to the committee in the past, but was especially pertinent to Reasoner's recent role in planning and developing the Michigan pool arrangements (see Chapter Thirteen).

The methods of the committee, developed from both theory and experience, were much more sophisticated than the simple projection of growth trends. The Company's own study of trends was supplemented by national and professional business statistics, and by a most intensive effort to forecast the effect on the Company of known changes, either in the local business climate or in the potential demand resulting from a customer's decision.

The committee recognized the possibility of recession and the reversal of the trends. And the long-range experience of the com-

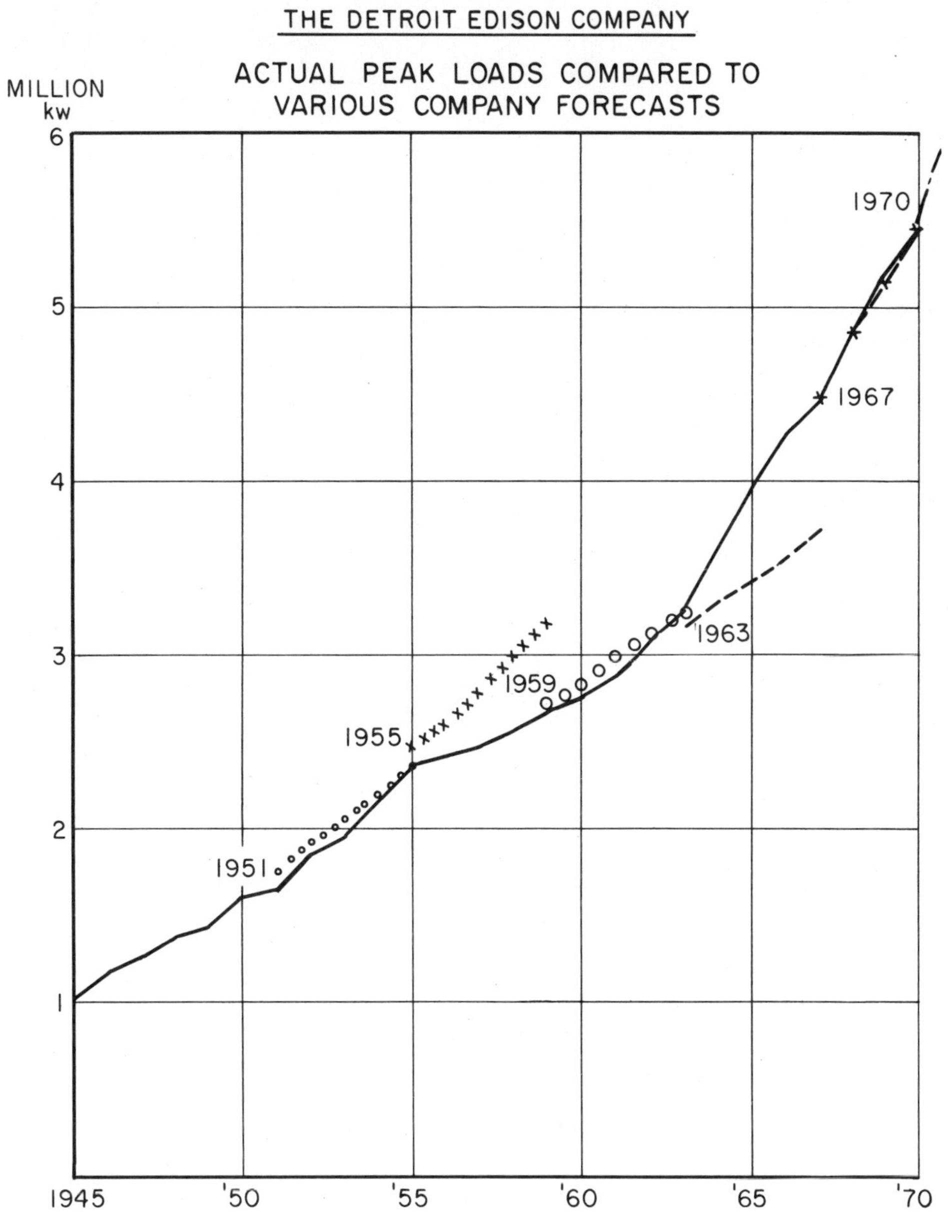

mittee itself justified their caution; economic change came suddenly and with small warning. For example, in the years 1952–56, the forecast increase of 8.3 percent had been close to the actual increase of 8 percent; the committee had forecast a continuation of this rate of increase, only to have the average increase for those five years (1958–62) stand at 4.5 percent.

In 1963, the committee used this figure as the basis for a forecast, together with its collective information, but the demand that year

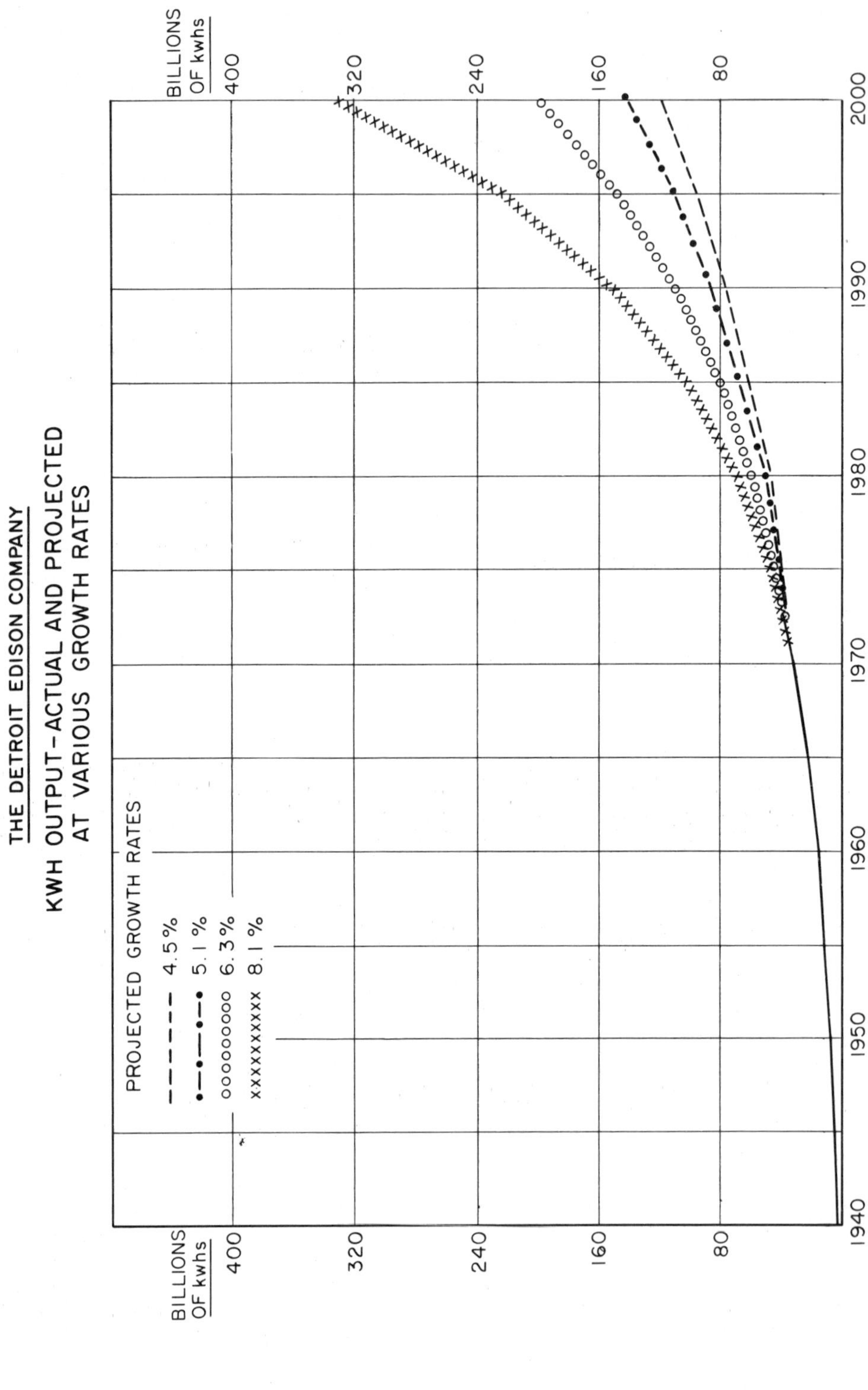
THE DETROIT EDISON COMPANY
KWH OUTPUT-ACTUAL AND PROJECTED
AT VARIOUS GROWTH RATES
PROJECTED GROWTH RATES
4.5%
5.1%
6.3%
8.1%
BILLIONS OF kwhs
400
320
240
160
80
1940
1950
1960
1970
1980
1990
2000

jumped by 8.1 percent. The next year, 1965, the forecast figures were revised upward to a cautious 5.1 percent; in 1967 to 6.3, and then for 1969 and 70 to 8.1 percent again. In each year, the percentage figure was used in the forecast for the five year period which would follow, and in each case, the estimate was low.

The effort to project beyond the year 1969 might lead to the use of some awesome figures, but there was no alternative to such an attempt. The time lag between decision to construct and the completion of construction was tending constantly to increase, and by 1969 it was essential that firm plans look as far ahead as 1976, a seven year span.

By 1975, a peak load of 10.5 million kw would develop, up from 1969's peak of 6 million kw. This would require an average annual increase in capability, over the seven years, half as great as the total capability of the Company in the days following World War II. To build these plants, lines and stations, the Company must invest over the seven years $2.2 billion. The capital expenditure would average one million dollars for every working day for the seven years ahead.

For the immediate future, the demands and costs could be calculated with some assurance. There was less certainty for the longer range, but the Load Study Committee used its next information to project trends into the second decade. In such a calculation, there were factors beyond either knowledge or control which defied confident forecast. On this information, modified as it would be by temperment and experience, the men in management must plan the future of the Company and of the communities it serves. It was an exercise which demanded cool courage, and a willingness to act.

CHAPTER THIRTEEN

INTEGRATION AND THE ALTERED PATTERN 1962 - 1975

Proposals for Regional Security / Experience with Interconnections / The Creation of the Michigan Pool / Hydro-Electric and Federal Regulation / The Blackout of 1965 / The System and the EHV Lines / The Control Center and the Pool / The System of Regional Agreements.

THE COMPANY, IN SPITE OF CHANGES AND VAST GROWTH, WAS STILL basically oriented about the loop system which had been its technological triumph in the twenties. In its creation that loop, which in its simplest form had linked Marysville and Trenton Channel to Delray and Conners Creek, had established the form and structure of the Company.

By the decade of the sixties, both the system and the techniques associated with it were outgrown. The existence of the extra high voltage lines, established in the Company at 345,000 volts, the altered economy in transmission of electricity and the sophisticated systems of control which were available, all combined to make possible a reorientation of the Company, and the rethinking of its organizational pattern. New techniques invited innovative men inside the Company and out to break with tradition.[1] In the new scale, the regional relationships were to become especially important.

In 1961, the Edison Electric Institute published an important report which was concerned with the problems of intercompany relations, and especially with the importance of such connections in maintaining reliability. The chairman of the EEI committee which prepared the report was Walker Cisler, whose concern with the problems common to the industry had been repeatedly demonstrated. The EEI was the active and responsible agency representing the privately owned power systems of the nation, and its committee was a prestigious one.

The report carried the recommendation for the broad extension of the system of interconnections, to the end that both regional security be achieved and maximum economy be attained. The experience of Detroit Edison and Consumers Power Company over the previous decades, supplemented by that of other companies and areas, was developed into a program recommended for the private companies of America.

In 1964, the Federal Power Commission issued a series of volumes, collectively called the Federal Power Survey. Members of the industry insisted that the acceptable features of this survey were in large measure duplicative of the EEI report, and that in any case it must not be accepted as a blueprint for the future. The survey included proposals for an elaborate governmental interregional transmission system, linking present and potential federally developed power sources with areas of increasing demand.

To men in the industry, the import of the survey seemed to be political rather than industrial. It used the existence of technological change and innovation to justify a very great extension of governmental responsibility. Its system, as suggested, seemed strikingly like the British so-called grid system, which had been the instrument for socializing the industry there. The survey made promises of savings and of price reductions which were economically unrealistic and could only be classed as propaganda. For the moment, however, there was no money for an elaborate nationwide public program of this kind.

The balance of the survey emphasized, as had the earlier EEI report, the advantages of interconnected regional systems. Walker Cisler had in the meantime become president of EEI. Referring to the survey, he said that there were things in it with which he would disagree in part or wholly, and that it was premature to assume that the final form of the power system had been reached or diagrammed, but that the industry could live with it.[2]

Both the report and the survey had stressed the long range importance of the interconnected regional operating groups, and it was estimated that 97 percent of private industry was already in one or another such group.[3]

Surplus capacity was a continuing economic problem for the utility planner. It was a waste except for the fact that it was also "reserve," the protection against sudden load increase or reduction of production by accident or breakdown. One of the very great advantages of the Company's 120,000 volt tower line had been that it allowed the reserve from all plants to be used as a common reserve. The increased security to each plant was equal to that which an additional turbine-generator would have given, and the cost was much less.

These facts were not new, nor was the effort to extend the advantage to unit systems. Alex Dow, in the twenties, had dreamed of interconnections which would link Detroit, western Michigan, upper Ohio, Indiana and even Chicago, for mutual protection. Insull in Chicago even tried to estimate whether the time difference of one hour, with its attendant divergence in the daily peak, would justify the cost of mutual exchange of power between the cities. Dow's proposal began an extended discussion, but for the moment brought only limited results. The first interconnection, established in 1928, was only between Detroit Edison and Consumers Power.

In the original contract and as restated as recently as 1942, this line had no function beyond provision for emergency reserve. The postwar contract (1949) showed clearly the results of experience in wartime pooling.[4] The new agreement was designed, in the words of the contract, to facilitate: a. "rendering mutual assistance during emergencies; b. affecting the maximum practicable economy in providing the electrical power requirements of each system."

This contract established conditions and the methods of price determination for three services. (1) One company could purchase a given amount of power from the other by prearrangement. (2) Power supplied as a result of an emergency became purchased power if the emergency demand extended beyond one half hour.

(3) Each company established a normal reserve adequate for its own system; if for any reason that reserve was reduced, a deficiency was established, and the company was required to "purchase" capacity, or at least to pay for the protection which the reserve of the other company had provided. The purpose of this contract clearly had gone well beyond the mere provision of security in emergency.

The economy of large scale production was obvious, but neither company alone could take full advantage of these economies without the penalty of immediate over-building. The correspondence between James Parker and Paul Thompson for Edison and Dan E. Karn, vice president of Consumers, indicates that construction programs were to be staggered, and the capacity temporarily available from a new machine was to be sold via the interconnection. Purchase and power would allow the buying company temporarily to postpone its planned construction.

The basic contract was dated January 15, 1949, with the later purchase agreements supplemental to this document. In the active, expansive period which followed, the purchase arrangement was of limited use; demand tended to grow in both companies and left each pressed to provide for its own needs.

As early as November, 1951, the addition of the third interchange (Whiting-Monroe) required modification. Provision was made for "in" and "out" meters at each interconnection, so that billing could be for the difference only. Interchange of power was becoming a normal part of the industry's practice. The 1949 agreement ran to 1957, and thereafter was renewable on a year to year basis. Changing conditions, several supplemental agreements, and above all the results of experience, ultimately made a new contract desirable.

The river boundary between Michigan and Ontario was a barrier in both the political and technical sense. Dow had once proposed entering the Canadian market, and he did explore methods by which he could reach customers there. His invasion of Canada was repulsed, and the Hydro-Electric Commission of Ontario moved to provide electric service on the eastern shore of the Detroit and the St. Clair rivers.

The main obstacle to interconnection between the two systems continued to be political. It eventually involved the Company's securing special legislative action so that it would not come under federal jurisdiction for international commerce (see Chapter Three).

The lines across the river were relatively easy to construct. The action of the Commission in shifting from 25 cycle to 60 cycle standard production removed the most serious technical difficulty. Between Delray and the Keith Generating station in Windsor, conductors and ground wires were anchored to a tower and then lifted to 302 foot crossing towers. The distance between them was 2,386 feet. Similar tower structures extended a line over the river opposite Marysville, also for substantially half a mile. The lines went into service August 31 and September 14, 1953.

The interconnection with Hydro raised several problems. Consumers Power was as concerned as was Edison about the threat of federal jurisdiction, and it was not protected in the special legislation Edison had secured. Elaborate precautions were written into the contract to protect Consumers from any kind of "interstate commerce."

Of equal concern was the possibility that Edison's new obligations might in some way reduce the capability of that Company to meet its commitments to Consumers. This worry was eliminated only when Hydro's management, in a letter of August 27, 1953, formally subordinated its claims to those of Consumers Power.

At the eastern end of this chain of systems, there were other problems. Hydro-Electric had not been an isolated system, even before the interconnection with Edison, for it had lines across the border to Niagara-Mohawk and its eastern connections. Common interests and problems led to meetings between representatives of the three companies and the Canadian Commission—even to the provision for weekly telephone conferences.

The interconnected group spread beyond the region. Interconnections from a remote New England group, chiefly small companies operating as a unit—led to the lines of the New York Authority, to Niagara-Mohawk, to Ontario Hydro-Electric, to Edison and on to Consumers. There were advantages and also dangers. The advantage came from pooling reserves and the efficient use of surplus power. The danger lay in the great extension of the area in which trouble could develop. The whole complex must be operated in phase.

So long as all went well, the automatic mechanisms could maintain the essential synchronism, even over so extensive an area of interconnection. But an overloaded machine, or a defective one, if it went only slightly out of phase, could produce trouble which would cascade through that system and potentially through these new and extensive interconnections to areas far from the trouble spot. In 1958, Edison moved to protect itself from that possibility by installing controls on all the Hydro-Electric interconnections, which would assure that an uncontrolled fault on either system could not cause an outage on the other.

On December 22, 1962, Detroit Edison and Consumers Power signed an important Pooling Agreement which was, in many ways, the logical conclusion of a generation of development. In the future, as rapidly as equipment could be put in place and plans developed, bulk power facilities of these two great systems would be planned and operated as one. It was a logical but imaginative extension of interconnections. Both companies viewed the move as significant for the Michigan area, and both widely publicized what the agreement did and did not do.[5]

This was not a merger of the two corporations, although this had been discussed, but probably never widely advocated. Neither company could ignore such forces as institutional pride and personal career satisfaction.

Another suggestion, probably more seriously considered, was that the purposes of the agreement could best be attained by the creation of a third corporation, the ownership of which should be associated totally with the ownership of Consumers and Edison. Whether a holding company or a jointly held company, this proposal proved to be complicated both in law and administration.

The agreement, involved though it was, was in principle simple in the extreme. In administration and financial matters, the integrity of the two companies was complete. In operational matters, involving current supply of bulk power, transmission, and planning for future growth, the entire two systems were to be operated as one.[6]

The 1962 agreement between The Detroit Edison Company and the Consumers Power Company was the foundation on which planning for the future must rest. But it was accepted with confidence because for a decade, or even since the war itself, the record shows developing intercompany cooperation and at the same time improvement of essential techniques. The opportunities thus

The interconnection from St. Clair to the Ontario Hydro-electric System. A helicopter carried a line over the river.

The Control Center, located near Ann Arbor, Michigan, is the heart of the Michigan pool and its interconnecting systems.

opened to scientists and planners promised to make the agreement a landmark both for the companies and for the state.

The new contract, for example, provided for two-man committees on Planning and on Scheduling. These two committees had existed through the decade of the fifties, much of the time with Howard Seelye representing the Company in planning and H. C. Reasoner in scheduling. The inclusion of the intercompany committees in the written contract indicates how vital and how successful these committees had been. Reasoner continued to be involved in these matters when he became assistant director of the Planning and Project Engineering department.

The agreement forecast exciting developments. The earlier decision to move to 345,000 volt extra high voltage lines was reconsidered and then reconfirmed. The potential in this electric transmission system would change the accepted boundaries. There were talks and then studies of the possibility of lines to the south, in spite of the political tradition which discouraged suggestions of interstate operations. With the potential demand of the whole state to consider, it was possible to revive also the discussion of giant turbine-generators, and to make discussions of such things as pumped storage something more than idle speculation.[7]

The new administration which took responsibility in Washington in 1961 named a new chairman of the Federal Power Commission, Joseph Swidler. His early experience in public life had been in the New Deal period, and he had had extensive responsibilities in TVA, but at the time of the appointment he was in private practice in law. In common with the industry, his concern was with the assurance of reliability, but his background would lead him to seek it through the interjection of governmental decision or even intervention.[8]

In this situation, the independence of Detroit Edison and Consumers from federal control would seem to the new chairman to be irregular, to say the least, and might even be viewed as a kind of defiance. Ten years before, counsel for the Commission had insisted to the congressional committee that if the Company was not in law subject to federal control it ought to be; and this mood certainly had continuing staff expression.[9]

In 1963, Commissioner Swidler approached Vice President Lyndon B. Johnson on the possibility of the repeal of the amendment of August, 1953, which had granted exemption from Commission control.[10] In September of that year the Commission published a list of companies under its jurisdiction, and included Edison and Consumers Power. The companies denied this jurisdiction, and the Commission itself was dubious enough to sponsor legislation, which was not passed, which would have given it control in international transactions of utilities.[11]

This proposal of legislation led to hearings before a senatorial committee, of which Senator Magnuson was chairman. The related correspondence gave Swidler an opportunity to place his arguments on the record. Isolated operations, such as the Michigan pool, he held to be uneconomic and impractical, even impossible. Thus, through its Hydro-Electric system connections to Niagara-Mohawk, Edison was itself a part of this extensive interknit system, and was by this fact a part of interstate commerce.[12] Cisler's response was that the system had not changed since Congress had given the exemption—and neither had the constitution or the law.

The argument that Edison was in fact in interstate commerce, because of its link with Hydro-Electric, seemed to the Company at best legalistic and at worst absurd. Edison supplied power to Hydro only when it was deficient in power, and in that condition Hydro would not be supplying anyone else. With Hydro's full cooperation, the Company moved to prove its disassociation from Niagara and the power supply there. A complicated technical system of controls, meters and timing devices was installed, to assure Washington that even emergency help was denied interstate.

On January 3, 1964, and again in April, the Company received letters asserting federal jurisdiction and each time courteously denied the claim.[13] For several years, the Commission had ordered the Company to file reports. The Company provided the information requested, each time being careful to indicate that it was information freely supplied and not a response to an "order."

By odd circumstance, the incident which brought all this to issue was unrelated to the central controversy and was unforeseen. Hydro-Electric was building a new plant just across the river from St. Clair. The planning group representing Michigan and Hydro-Electric recognized the advantage of interconnecting these two plants, and asked the FPC for permission to install a third international interconnection.

The filing late in 1964 served to reawaken the discussion. The Commission ordered (February 4, 1965) that a docket be established and a hearing scheduled, and in the meantime staff members moved in to investigate the Company, and also Hydro-Electric, Niagara and Consumers. The Michigan companies granted access to such information as was relative only to the issue of interstate shipment. Hydro-Electric, which could claim to be an arm of the Ontario government, was a bit stiff about this proposed investigation by the agency of the United States, whatever its purpose.[14] The scheduled hearing was postponed and rescheduled for June, with formal written pleas required. The Company determined to withdraw the application, whereupon (July, 1965) the Federal Power Commission ordered proceedings to determine jurisdiction.[15]

Swidler had stated his case in the order for the hearing. The issue was whether or not the interchange, in the form for which application was being made, would in the language of the original law "impede or tend to impede the coordination in the public interest of facilities subject to the jurisdiction of the Commission." The argument was simple. If the flow of power was to New York, then it was interstate. On the other hand, if the free flow was being "impeded"—and that certainly had been Edison's avowed intention—then the interference in itself established jurisdiction, and was "inconsistent with the public interest."[16] It was a neatly contrived dilemma.[17]

The Northeast Blackout of November 9–10, 1965, was a traumatic experience to the electric utility industry. Comedians could exploit the situation to the full and did, and political defenders of the public interest might threaten and command, but to the men involved and bearing responsibility it was a situation incredibly and utterly shocking. Since the Federal Commission and the several state regulatory agencies could not escape at least a share of the responsibility, they, along with the men of the industry, found it a humbling experience.

The event itself can be briefly described. At 5:16 p.m. on November 9, under peak load conditions, a protective relay device tripped on a 230,000 volt line supplying Toronto from the Sir Adam Beck plant at Niagara, Ontario. There was no malfunction of equipment or operation; the relay was set to cut out safely below the danger point. The transfer of the load of this line to the four others which operated parallel with it overloaded them, and promptly triggered their protective equipment, and they cut also.

These lines were carrying 1.5 million kw at the moment of separation, and this power sought outlet over the lines of the Power Authority of New York, already carrying 1.5 million kw and at capacity. The power surge of this magnitude overloaded the transmission facilities and controls. The wild swings from dangerously excessive capacity to inadequacy sent trouble cascading not only over the Canadian system where it began, but on to the interconnected systems and further separation occurred. In this situation the local areas were thrown onto their own production capacities which were inadequate, and large areas were unable to maintain themselves. By 5:30, virtually all generation in the entire northeast had come to an incredible stop, and in many cases power was lacking by which it could be started up again.

The story of the struggle of a system bereft of power even to operate supplemental equipment is, fortunately, not a part of the history of Detroit Edison. The troubles poured across the entire width of Ontario, and that area was blacked out. At the Edison connections, there were four safety devices, properly installed, correctly set and in working order. In one and two-thirds seconds, as the timing devices showed, the Hydro-Electric interconnection was cut and Edison was safe. Indeed, almost at once, the flow of power from Edison began supplying Hydro and its customers as far east as London, Ontario.

Suddenly, in the wake of the common disaster, both the Commission and the Company were led to reconsider the matters at issue between them, including the request for the third interconnection with Hydro. Even the question of jurisdiction and federal responsibility seemed relatively less critical or less significant to the industry. The blackout crisis precipitated decisions which probably would have been made anyway, but which now came more promptly and more easily.

The Company accepted, as an ultimate certainty, the jurisdiction of the Federal Power Commission. As late as in 1964, the planning committee of the Michigan pool had stated that the advantages of interconnection to the south did not justify the bother and expense which federal responsibility would require. In 1965, that judgment was reversed; when and as the 345 kv interconnections with Hydro were completed and the lines built between Edison and Consumers as planned, both reliability and economy urged the completion of similar connections to the strong electrical systems of Ohio and Indiana. When that connection took place, the Company would then accept its definition as an interstate company.

On December 20, 1965, a carefully drawn letter was sent to the board of directors which showed how fast events had moved. In the month since the blackout, staff discussions in Edison and with Consumers had matured plans long in development. The board of Consumers Power would meet on December 27; since Edison's board would not meet before that, he wrote to inform its members of agreements approved on staff level, which would be presented to Consumers Power and would come to the Edison board in the next meeting. He would then suggest that the board authorize the agreements for and the construction of the transmission and switchgear facilities essential to the operation of the southern interchange by January, 1970. The board was to meet in Detroit on January 17, 1966.

The balance of the arrangements were simple, especially in the new atmosphere. Paul McQuillen, New York counsel, who often represented Edison in Washington, and Harvey Fischer, Michigan counsel, negotiated a settlement with the newly appointed leadership of the Federal Power Commission.[18] Federal jurisdiction would be accepted without further debate or hearings. The Commission would drop its claim to past jurisdiction, would accept the date of July 1, 1969, when the first of the southern interconnections would become operative, and in the meantime would permit the installation of the third interconnection with Hydro-Electric in time for it to be operative in 1966.

There had been some earlier diplomatic gestures by the Company, especially on the part of Walker Cisler. He had accepted Swidler's invitation to the Washington dinner incident to the presentation of the Survey by the Commission. As president of the EET he had been conciliatory and generous in the discussion of that Survey. Soon after, he took two public power men with him on a tour of Romania, as he felt both the public and the private sectors of the industry would be better for the exchange. It would also be helpful in reminding the Commission that private power was committed to serving the public interest.

By April, 1966, the press carried both the story of the new Canadian interconnection and the first announcement of the even more extensive plans for connections to the south. The announcement covered the matter of federal jurisdiction, but tried to emphasize continued state and local responsibility. It was a curious end to a conflict which had cost Company men many troubled hours.

The years 1962 to 1965 were a time of extraordinary change in the conditions under which the Company would operate. The remarkable pooling agreement with Consumers, the removal of the limitations under which Edison worked with Hydro and, above all, the freedom allowed by the interconnections across state lines to the south combined to create a substantially new situation. Both production and transmission responded to the opportunity.

The technical problem can be restated in simple terms. Economy demanded that the utilities move to vastly larger machines than they had used. The risk of outage of so large a machine brought to mind the old rule of thumb that no single machine should exceed 5 percent of peak demand; usually it had required that the reserve available exceed the largest single machine. No matter how the rule was stated, escape from the limitations on size could only come with the development of a transmission system which could assure power over sufficient distances and in adequate amount.

The so-called presidential letter from the FPC, giving permission for the Lambton-St. Clair third interconnection was dated March 1, 1966. The steel for the towers had been ordered in advance, and it was planned to have the lines in place by the end of September of that year. At that time, the Company already had under construction some 90 miles of additional 345 kv lines. It was the beginning of the newly designed system.

Basic to that system was the utilization of the new 345,000 volt transmission lines as a kind of backbone for the integrated two companies. From roughly the location of Wayne, Michigan, the line stretched west across the state. At the west and east sides of the state, the vertical lines extended from the state line northward.[19]

The important new element was that these towers did not stop at the state line. From the spot where the Hydro-Electric intercon-

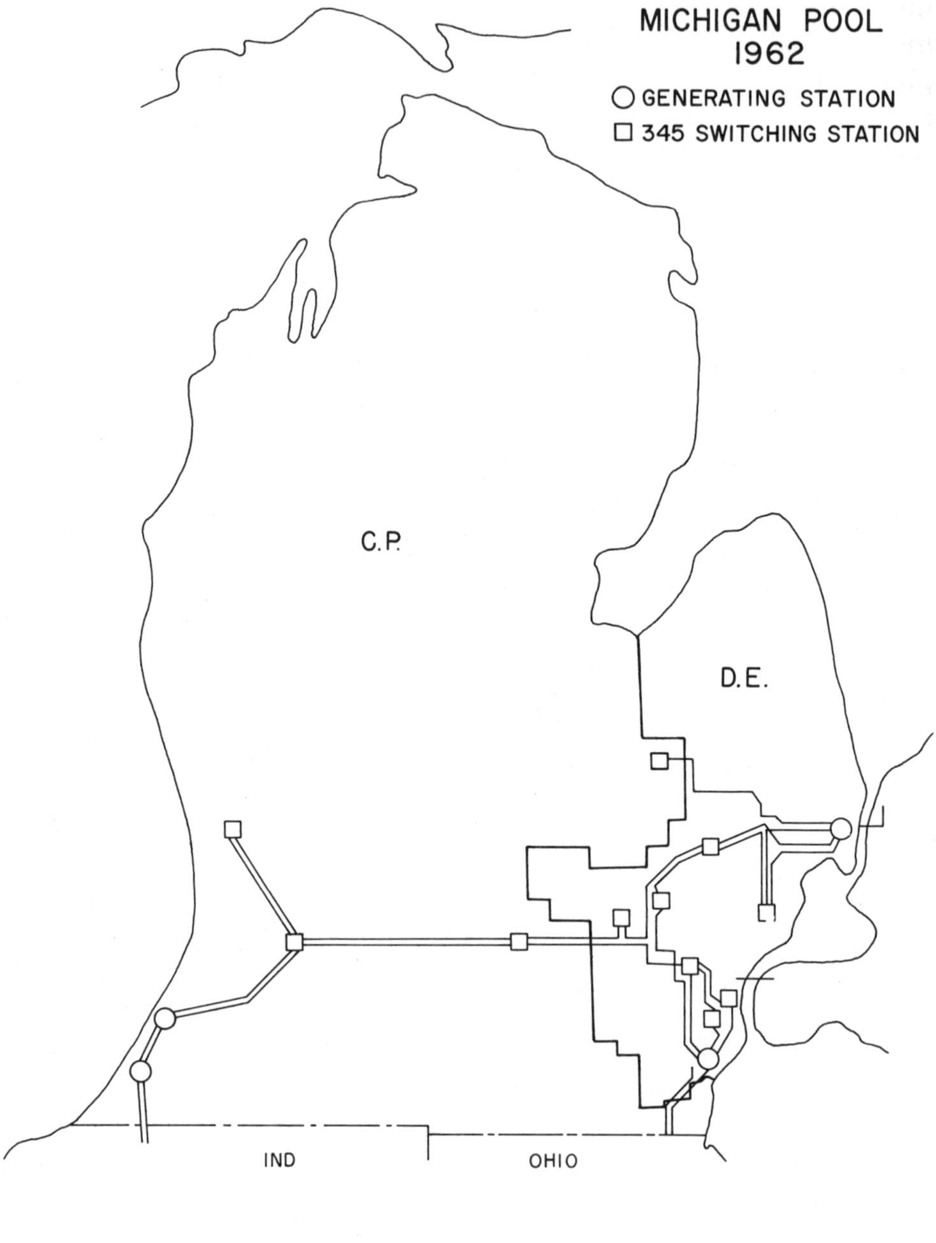
MICHIGAN POOL
1962
GENERATING STATION
345 SWITCHING STATION
C.P.
D.E.
IND
OHIO

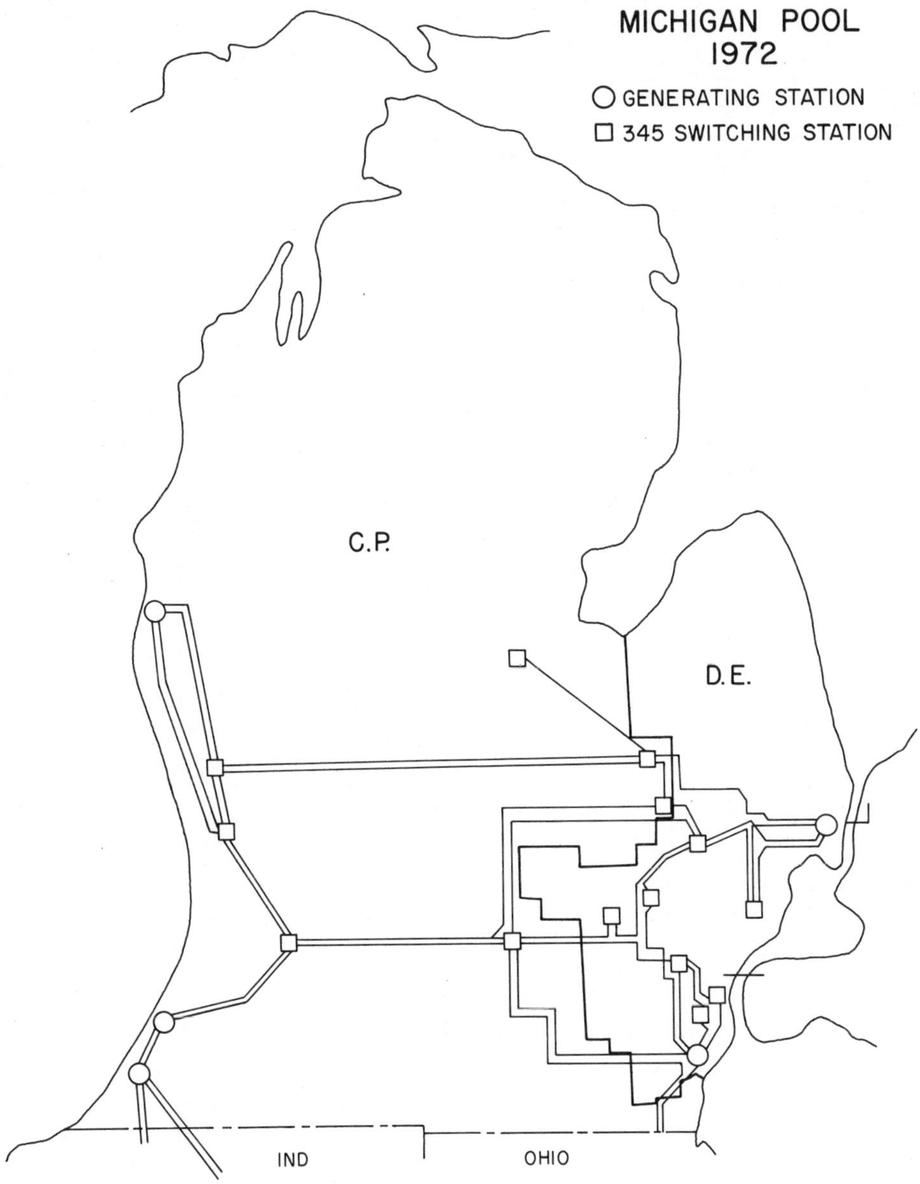
MICHIGAN POOL
1972
GENERATING STATION
345 SWITCHING STATION
C.P.
D.E.
IND
OHIO

nection could be most conveniently linked, the tower line extended southward, through franchised territory of Detroit Edison, Consumers Power, Toledo Edison and American Electric Power, to near Lima, Ohio, where it connected with the 345,000 volt line of American Electric. On the west, the shorter distance from Consumers Power line also involved several companies before it reached the same ehv grid of American Electric Power in western Indiana.

This construction of roughly 175 miles of high voltage line, when completed and interconnected with the already existing systems in Indiana and Ohio created a new loop. The ehv lines of American Electric Power Company closed the distances between the bottom extensions. The seven utilities thus connected served roughly twenty million people, and had a generating capacity which by 1970 would exceed thirty million kw. To each of these utilities, the knowledge that support could come from either of two ways gave assurance of high reliability.

However, these matters were not left to chance. The system was large, but largeness alone was no guarantee of reliability. The group carried out a series of contingency studies in 1968; testers threw multiple disasters at a simulated system to see what the consequences would be. The earlier experience of American Electric Power with 345,000 volt lines was confirmed; the strength of the system minimized or eliminated the separations, and localized the area of the assumed disturbance.

In anticipation of increased demand, the companies in the pool planned also to close the distance between the upper extensions, with a line roughly from Flint to Grand Rapids, and thus to create another ehv loop. Plans included extension of the west line northward to the projected Ludington development, and the east line to the important Bay City-Saginaw area, all at 345,000 volts.

The first section of the interstate 345,000 volt line was activated July 1, 1969. The entire development was scheduled to follow over half a decade.[20] Men in planning and in production moved to take technological advantage of the new Company scale. The changes in organization and administration were as innovative as were the changes in transmission.

The decision contained in the agreement of 1962, to operate the two systems as one, challenged the planning groups to fully utilize the available techniques and resources. Out of the study came plans for an elaborate central control system which would support and in part supercede the separate company operations.

Each of the major companies had developed an integrated system of control. From the centers in Jackson and in Detroit, a network of communication and control reached every part of the system, directed normal operations and reacted to emergency in line and plant. Because of the importance of this operation, the equipment was repeatedly updated, and on several occasions had been substantially rebuilt.

The decision to operate the pool as a single system called for a new system of control. The Control Center was built by Edison and is in Edison territory. Operating costs are shared by the two companies, and the personnel is furnished by both. The building itself is simple in appearance. It was built of poured concrete, sand blasted for an attractive finish, and located on spacious grounds in Scio township, near Ann Arbor. It is two-story, with offices and public rooms on the upper level, and a surprisingly small working area on the level below. Despite the importance of the center there is little for a vistor to see; it is a quiet operation with no quality of drama. However, the Public Relations departments of the two utility companies have designed an informational exhibit for the visitors gallery in the building.

What the center does is more easily explained than how it does it. When it went into service the two companies had in operation approximately twenty major power stations containing some sixty-seven machines. Each of these machines can be individually operated from the center and, from each a continuing and instantaneous report shows how much power the machine is producing, and how far that figure deviates from the established most efficient production level for that machine. Responding instantly to any change in demand in either system or in any of the satellites, the computer in control decides where to cut, or where to find the most economical energy. If power can be secured, over interconnection, at less cost from the outside sources than from within Michigan, that decision is automatically made—and the amount of purchased power recorded. Eventually, the computers will add up the energy purchased, subtract the power sold over that interconnection, and figure the bill.[21]

The original intention had been to complete the center sometime before the end of 1968, in order that the staff might eliminate the inevitable "bugs" of so complicated a system. Delay in construction postponed the operation for some months, but the center, when it became operative, handled the whole complex with minimal difficulty.

The center was the logical development of the agreement of 1962. For both companies, it symbolized the search for economies from pooled reserves, the mass production of electrical energy, and the jointly planned extensions which the agreement had anticipated.

The conception of regional organization for electrical utilities was in no sense new. The very real economies which could be obtained from intercompany transactions had been the original justification of the holding companies among utilities in the 1920's. Financial excesses ultimately discredited these organizations, and the problem thereafter was to find some acceptable method by which to secure these operating advantages. Short of actual merger, the most complete integration could be illustrated by the Michigan pool, and the type of close coordination which guided Edison and Consumers Power.

For several years before interstate connections were closed, the appropriate officers of the two companies had been in consultation with their counterparts below the state line. Out of these discussions had come a group of intercompany agreements collectively known as the MIIO agreements, obviously referring to the states of Michigan, Illinois, Indiana and Ohio. On the west, Chicago's Commonwealth Edison was included, on a somewhat restricted basis, since that company tended to associate itself with the region west and north of the city.

Intercompany assistance in emergency could be assumed; this was so important that in the MIIO agreements each company agreed to go to all practical lengths in providing assistance in order to prevent disaster in a neighboring system. The recent experience with the blackout of 1965 was obviously reflected in this rather drastic provision. Fortunately, such a contingency is a rare experience. The intercompany arrangements for pooled reserves and adequate transmission were designed to prevent this kind of crisis.[22]

Beyond emergency, important though it was, the associates were concerned with normal operations and especially with bulk power. Coordinated joint planning and the interchange of information supplemented the planning of the individual companies. Arrangements arrived at on staff level provided for planned increases of energy capability for the whole group, and for the distribution of that power by purchase or exchange as required by member companies.

The MIIO arrangements with individual companies dated from 1966 and 1967. The Michigan pool was in fact joining an earlier informal group of associates which clustered about American Electric Power, and which was strongly influenced by Philip Sporn, its long-time president. American was a holding company, with subordinate and sometimes isolated companies, the needs of which had given experience in intercompany operation. The American Electric Power Company was slightly larger than the Michigan pool in peak capability.

The new dimensions added by the interstate and international interconnections altered day to day practices in operation. The purchase of bulk power, the contracting for future delivery of a given amount for a given length of time, even the search among a number of companies for one capable of meeting a purchase request—these all became commonplace and routine in an industry which a dozen years before would have found them both technically and administratively impossible.

Even more dramatic was the development of the several regional organizations, intended to reinforce regional reliability arrangements.[23] Sporn and American Electric had worked with the Federal Power Commission from its inception, perhaps anticipating the unavoidable. In any case, the Commission's concern with the problems of reliability, as disclosed in the 1964 *Survey,* and as emphasized after the blackout, coincided totally with both Sporn's experience and his trend of thought. The association of companies on a regional basis would allow them to provide for themselves and to give each other that measure of protection and cooperation which might otherwise be imposed from the Commission.[24]

The East Central Area Reliability Coordination Agreement (ECAR) and the MIIO agreements seemed to overlap, but this was not important, for the two organizations had different purposes.

ECAR was neither a pool nor a planning organization, but a coordinating instrument. The organization document and agreement was approved January 14, 1967.

Quite frankly, its purpose was to prevent another blackout. Twenty-six electrical companies in eight east central states joined in the announced effort to achieve the highest degree of reliability in their bulk power supply. To accomplish this goal, they proposed to adopt uniform procedures in planning and operating their systems. The plans of each company for new generation or transmission, together with existing reserves and procedures, would be subjected to tests which would demonstrate their adequacy and reliability.

The organization of ECAR is an indication of its importance as viewed by the founding companies. A key executive officer from each company served on the board; from Edison, it was Walker Cisler himself, with Ed George as his alternate. The operating heart of the whole organization, as it was described by its president, was the Coordinating Review Board, supported by some half dozen specialist committees. Harold Reasoner was the natural Edison appointee to this Board, as he had been a key figure in the development of the Michigan pool, and in the MIIO agreements.

Even ECAR, though it covered in part some eight states, found itself with boundaries, and with concern for what lay beyond. The problems of other regions were in some degree unique, but the need for some kind of communication had led or would lead to organization similar to ECAR but quite diverse in detail and even in scope.

The National Electrical Reliability Council was created, after preliminary conferences, on June 11, 1968. The report of the membership, January, 1969, shows twelve regional organizations, with TVA counted as one of the twelve. This organization saw itself as the one to provide independent review to interregional matters referred to it by the regional organizations, to provide the Federal Power Commission and other federal organizations with information as requested, and thus to speak, when appropriate, for the industry. The organization was, quite frankly, the choice of the industry as against one proposed by the Commission to perform this advisory function.[25]

For the Company, then, the Control Center in the Michigan pool acquired increasing importance. Teletype continuously gave information from all of ECAR itself, together with significant items from its contiguous neighbors. By leased wire, the Center was in continuing connection with Canton, Ohio, and American Electric Power's headquarters there. In the same way, a leased wire gave information from Hydro-Electric and the companies in the eastern United States associated in CANUSE.

An illustration from the first summer of its operation will indicate the importance of this information and the response. ECAR received and transmitted word of an emergency in PJM, the operating pool in Pennsylvania, New Jersey and Maryland; not one but several machines were out, and in the heightened summer demand they must have help. With other companies, Edison was able to make its contribution; energy supplied by the Company from Michigan served Philadelphia and Washington over the lines of more than a half dozen interconnected companies, and through as many states. There was no blackout.

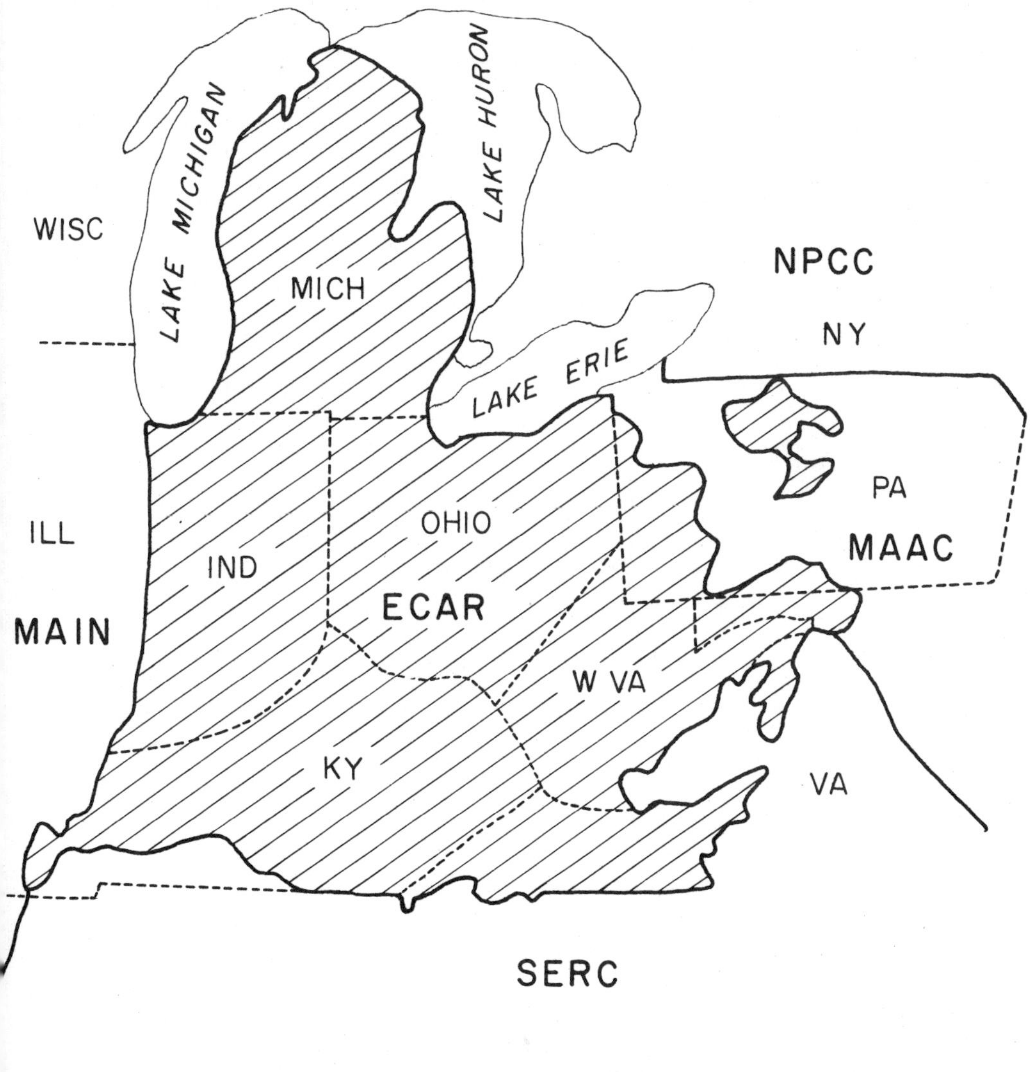

CHAPTER FOURTEEN

THE PRODUCTION OF POWER 1962-1975

The Program of the Pool / Improvisation and the Peakers / Monroe; the Completion of a Series / Ludington; the Incredible Project / Fermi No. 2; the More Conventional Atom.

DETROIT EDISON COULD IN 1962 POINT WITH PRIDE TO TWO QUITE NEW and modern plants. Both had been built in the fifties, following the new leadership. Each plant represented the advanced technology of that decade. To complete the planned complement of these two plants, three machines rated at 300,000–325,000 kw were installed in 1958, 1959, and 1961. They were the largest in the system, and were among the largest machines of the industry. In the test of use, they proved to be efficient, economical, and with some exception, reliable.

The slow pace of recovery from the recession of the late fifties tended to color all plans and estimates for future years. For Edison, unhappily, there appeared to be no pressing need for construction. At the close of 1961, the peak demand was 2.9 million kw and the capability 4.0 million. Even when the sales and demand mounted statistically in 1962 and 1963, management tended to discount the implication. By 1965, however, the trend was unmistakable.[1]

Under the pooling agreement of 1962, the two Michigan companies were committed now to the longer point of view, and the planners were required to develop a coordinated program which related prospective customer demand to productive capability and financial requirement.

The result was a kind of master plan, covering the period 1965–75. There is a temptation to refer to it as a timetable, except that is exactly what it was not. Time was the elastic element, the method by which adjustment could be made. Increase in demand must be accepted as inevitable. The two companies would assure themselves of adequate capability by constructing adequate facilities. They would adjust to changing circumstances either by accelerating or retarding the pace of the construction program.

The long-range program of the Michigan pool was obviously several years in development, during which time it was subjected to acceleration and to substantial increase in planned capacity. Soon after 1965, the planning groups were able to lay out a construction program which anticipated the decade and its growth, and allocated responsibility for construction. The table which follows represents the thinking as of 1968; later modifications were chiefly in time and to a degree in size.[2]

	Plant	*Company*	*Fuel*	*Capacity*
1967	Campbell #2	Consumers	C	385,000
1968	Trenton #9	Edison	C	519,000
1969	St. Clair #7	Edison	C	527,000
1970	Palisades #1	Consumers	N	700,000
1970	Monroe #1	Edison	C	800,000
1971	Monroe #2	Edison	C	800,000
1972	Monroe #3	Edison	C	800,000
1973	Monroe #4	Edison	C	800,000
1973	Ludington	Joint	PS	937,000
1974	Midland #1	Consumers	N	533,000
1974	Fermi #2	Edison	N	1,150,000
1975	Midland #2	Consumers	N	792,000

The first announcement of construction plans was by Consumers Power. In April, 1964, that company announced a new 385,000 kw unit to be added to its James H. Campbell plant on Lake Michigan west of Grand Rapids. According to the plans, this unit would be in production by April, 1967. For its first year of operations, Edison was assured of substantially half of the total production.

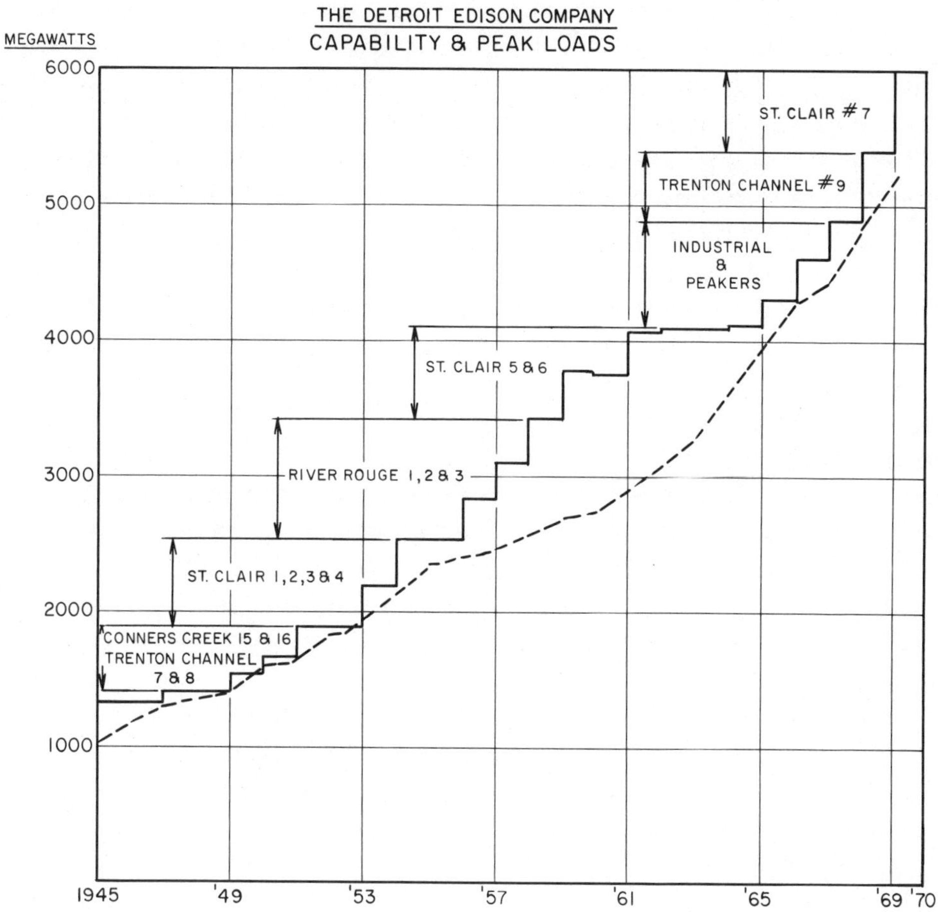

Very soon after Consumers' announcement of its first power pool unit, Edison announced its first pool unit to be installed in Trenton Channel as No. 9, and planned to be operative by January, 1968.[3] In January, 1965, the representatives of the two companies agreed that the Michigan pool would need another unit by 1969, and Edison agreed to construct it. This became St. Clair No. 7.

The situation indicated that the Company should move promptly with both projects. Invitations went to both General

Electric and Westinghouse to bid on the required machines. Cisler discussed with the board the situation which existed since Allis-Chalmers had withdrawn from the manufacture of this type of equipment. He thought it unfortunate to have the number of competitors for American business reduced to two. Thought was given to asking also for some bids from abroad, but that was abandoned because of the tightness of the manufacturing schedule.[4]

The contract for Trenton Channel No. 9 (519,000 kw) went to General Electric. The turbo-generator for St. Clair No. 7 was from Westinghouse, and was rated at 527,000 kw. Combustion Engineering was given the contract for the steam generators, which were substantially duplicate units. Boiler and machines were designed to operate at 2,400 psi (pounds per square inch) and at 1,000 degrees Fahrenheit heat and reheat.

In some regards, Trenton No. 9 and St. Clair No. 7 differed from the huge units to be built later, but the family resemblance is obvious. The characteristic physical form had already been evolved. The one machine and its steam generator occupied what was substantially a complete building, separate from the older structure on the site and technically independent. It was referred to as an extension, and offices, shops, and some other services were shared.

The dedication of Trenton Channel No. 9 occurred on February 8, 1968, when Harlan Hatcher, a member of the board, pressed the button which set the huge machine in motion. St. Clair No. 7, despite construction interruptions, was ready for dedication ceremonies on April 18, 1969.

This was an emphatic reaction of the Company to the obligations incurred in the pool, and to the altered situation in the economy. These two machines had a combined capacity which exceeded 25 per cent of the Company's total when they were planned in 1964–5. The steady flow of energy from these machines gave the Company at least a temporary relief from the worry over low margins of capability.

In 1964 and 1965, plans were being matured which would meet the Company's demand, hopefully, in the year 1969 or 1970 and beyond. In the meantime, present mounting demand must be served. The Company suffered the not unfamiliar troubles which come from a conflict between long-range purposes and immediate necessity. For a number of years beyond 1965, improvising became a kind of way of life, and real emergencies were avoided only with difficulty.

The demand, as measured by the successive years' peak, was rising sharply. In 1964, the increase over the preceding year was 335,000 kw. The next year it increased 360,000 kw, and for the first time in a decade, the announced capability of the Company was exceeded by the demand. In his letter to the board in July, 1965, Cisler called attention to the fact, and suggested immediate action before the summer of 1966, lest there be a real disaster. The solution for the problem as the chairman saw it, was the use of "peaking units." These were small, sometimes even mobile, units powered by diesel or oil fired turbines; most importantly, they could be obtained in much less time than more conventional equipment.[5]

In the interval, however, two other sources of power, more significant than the "peaker," had become available. Obviously they were not a part of a program of improvising, though the effort to hasten their availability is an indication of concern with that problem.

A number of years before (1954), the Company acquired from private owners and from the state some land, part of it submerged, near Harbor Beach, in the northern part of the Thumb. In February, 1965, Don Kigar, in his presidential address to the Power Club, announced the decision to proceed with plant construction there. It would be a plant of relatively modest size, which might be rapidly constructed and that was important.[6] There were several technical advantages in such a plant and location. It would strengthen the reliability of Edison service, often endangered by the effect of storms on the long lines from the south. It would meet the rising demand which population and industrial growth was producing and which would no doubt continue.

Dredging and filling began on the site even before contract bids were formally submitted in 1965. Construction began the next year. In May of 1967, the Company was trying to push forward the start up of the plant from June to January, 1968, but with modest results; the start up was only six weeks ahead of schedule, but in the summer of 1968 that was important.

As the Company was coming to view such matters, the Harbor Beach plant was small. The one Stahl turbine generator was of

Swedish manufacture, shipped via the *Mai Rogue* to arrive directly at the Detroit Terminal, September 21, 1967.[7] There had been a good deal of talk about foreign purchase, but this was the first relatively large machine imported. On the line, after the formal dedication, the machine tested at 114,000 kw.

Even before the Harbor Beach plant came on the line, a measure of relief came from another, rather unusual, source. Edison owned and would operate the electric generating equipment of the Fermi nuclear plant. Fermi was frankly developmental, and by 1966 it was clear that the installation of the turbine generator had been premature (Chapter Sixteen).

Without steam from the reactor complex, the 150,000 kw generation equipment was idle, at a time when it was most needed. In 1965, an oil fired boiler was installed to make provisional use of the machine. By February 15, 1966, when the plant came on the line, it was known that the 150,000 kw unit would be critically important in the impending summer. No one, however, expected a new peak as early as February 22, just a week after start-up.[8]

The first of the peaking units was installed on the grounds of the Fermi plant not far from Monroe. This unit consisted of four machines and was designed for unattended operation. On April 29, 1966, the four turbines automatically fired, accelerated, synchronized with the system and loaded properly. The package unit of four machines was made by General Electric and provided 62,000 kw capability.

In 1966, including the installation of the package at Fermi, peaking units were at four locations. One was a package of five diesel generators totalling 10,000 kw and installed at Dayton Station; the others at Northeast and Superior were similar to the units at Fermi. The package for Hancock was ordered from Westinghouse at the same time, but was delayed in manufacture. Including the first unit at Hancock, delivered in 1967, these peaking units totalled 276,000 kw capability.

The most significant function of the peaking unit was, of course, to solve the problem of the peak. These combustion turbines and diesel units were quite expensive to operate, as opposed to the high efficiency of conventional generation. This sharply higher fuel cost per kwhr produced was only in part offset by the fact that the machines were unattended. Most important, however, was the fact that the system operated at its peak only a limited number of hours per year; for this limited time the operation of the conventional plant would be most expensive. The peak in demand must be met, and for this limited purpose the peaking units were both modest in cost, and available.[9]

There were some other advantages which would follow the installation of the peaking units. The units could be located in areas and on lines which stood in need of support, either because of increased demand or because of irregularity in load. The characteristic which proved most immediately important, however, was the fact that they were substantially autonomous.

At the time of the blackout of 1965, a number of eastern plants shut down by the operation of their safety mechanisms, found that they were helpless to start again without power from the outside, which was unavailable. The capacity for a "black startup" became a safety factor not only for a system as a whole, but for each individual plant. After the blackout, the Company promptly examined its situation. Small peaking units, to a total of about 50,000 kw, were added to the system late in 1967 and early in 1968. Their location, at River Rouge, Harbor Beach, St. Clair, and Trenton Channel (Slocum) indicates the places where this type of additional protection was necessary, and where the peakers could do a double duty.

The validity of the basic argument which justified the use of the peaker units depended upon the limits put on their hour use. When the first unit at Fermi was declared operational, the announced limit of expected operation was two hundred hours per year.[10] As contrasted with conventional machines, these peaking units had life expectations which might be as short as five thousand hours of operation; at two hundred hours per year, that limitation was not significant.

In 1968, the larger peaking units were on the line more than twice as long as theoretical economic use justified. Even after St. Clair No. 7 and Trenton No. 9 came on the line, the peakers were used far more than planners had assumed they would be. In spite of the increase in conventional production, in the summer of 1969 the peakers were in use at a rate which projected well over three times the two hundred hour ideal. The effect of the use of these high cost units for so extensive a period helped raise the fuel cost per million btu to the highest figure it had reached since 1958.

In 1959, the annual report announced the appointment of an Advanced Projects Committee. Its assignment was to study and report on new methods, or proposed new methods, in production or distribution of electricity. In 1961, the Edison Electric Institute organized a task force on advanced projects, to keep the industry as a whole informed. Harvey Wagner was chairman of both the local committee and of the task force. Late in that year, he and W. G. Meese, then director of engineering research, made an extensive tour of England, France, and Italy, discussing with scientists their proposals or experience in advanced procedures. They returned persuaded that American technology was not behind Europe.[11]

Announcements of possible new techniques were frequent; it was essential that they be assessed. The "little black box," the regenerative fuel cell, was practical but only at high costs; it produced electricity by chemical reaction. There was excitement about both solar and nuclear energy as a means of direct production of electricity. The so called "plasma" generator was promising, but demanded temperatures which were for the moment impractically high. Laboratory experiments created talk about the production of electricity directly by fusion; but despite encouragement and even subsidy, the revolutionary methods did not materialize. Edison's development would still involve either conventional hydro-power or thermal power, with the conventional generator attached.[12]

Edison had planned in 1964–65 its first two pool units, Trenton No. 9 and St. Clair No. 7. Beyond these major projects, the long-range plans called for Edison to construct two new power plants, and to share with Consumers both the costs and advantage of a third project.

The new Monroe plant on Lake Erie is in the extreme southern tip of Edison territory. When the 1,200 acres of land and marsh were purchased in 1953, the location could have been important. But by 1965, when early estimates of growth in population and energy demand had been exceeded and with the creation of the pool and the 345,000 volt transmission system, it was obvious that the plant built here would be a major one.

At the time of this decision the possible use of nuclear fuel was studied but rejected. The Monroe site, at the mouth of the river and directly beside the Monroe harbor facilities promised significant advantage in the transportation of coal, including alternative shipment by rail or by ship. The existence of the ship canal would aid in price negotiation with the railroad, even though the railroad was from the first the designated agent.[13]

In the preliminary Company meeting in May, 1966, it was agreed that the Monroe site would permit ultimate development to about 3.5 million kw. The assumption was that immediate construction would be of a first unit. On August 1, 1966, however, at a meeting with officials of Consumers, it was agreed that the pool would need this first unit by 1970, and the forecast called for a second by 1972. The decision was made jointly that this second unit would also be at Monroe, and every effort would be made to take advantage of the economies of duplicate construction on the same site.[14]

On August 15, 1968, orders were placed for equipment for the third unit, and on February 17, 1969, the fourth unit was ordered. The estimated five years of construction could be reduced a little in this ordered sequence, and Monroe No. 4 was scheduled for completion in the spring of 1973. These were pool units, and the output would be shared with Consumers, but it is not the less interesting that the developing plans changed from one unit to four, from 800,000 kw to 3,200,000 kw, between May, 1966, and February, 1969.

The units are substantially to be duplicates, both in form and in equipment. They are self-contained, except that each of the two pairs has a connecting structure to house some offices, services, and controls common to both. Each pair is connected to a single stack.

In the development of the plans for Monroe, the transition in form and size which was noted in Trenton Channel and St. Clair was further extended. The notion that there is a structure into which equipment is moved would not survive even a slight inspection. There is a roof and a kind of metallic weather screen on the sides, attached to ribs and supports. The courts, for tax purposes, permitted depreciation of this structure, at the rate of a machine rather than as a building, recognizing the realities of this technological creation.

The four units standing side by side will extend one fifth of a mile across the reclaimed area. The individual units, roughly 265 by 350 feet, are larger in area than a football field by a considerable margin, and will be 13 floors (210 feet) high. Inside this huge frame are two basic elements, plus the devices to make them function. One of these is the steam generator, and the other the turbine generator which translates steam into electric power. The purpose of the whole complex is to translate energy from the less usable form of coal to flexible and available electric energy.

Coal is reduced in ball mills to powder form, and then blown on super heated air into a combustion chamber which bears small resemblance to the old furnace. This form of combustion, which burns the coal in suspension, is as old in the Company as Trenton Channel, but was never before used on this scale. The combustion chamber is 45 by 80 feet, and 200 feet high, with temperatures as high as 2,000 degrees. At full capacity each of the four units burn three carloads of coal per hour, a total of 300 tons to produce 5.5 million pounds of steam per hour.

If the plans for fuel and fire were conventional, the steam generating equipment was not. Edison and the industry had carried pressures to 2,500 psi and temperatures approaching 1,100 degrees. There was no point in going to higher temperatures, for the costs of the special metals required more than offset known economies. The unique thing at Monroe for the Edison Company was the use of steam at supercritical pressure. The design of the steam generators was sharply modified to accomplish this feat, and they became the first such units in the Company.

The old image of a boiler as a tank of boiling water and steam, both under pressure, from which steam could be drawn for use, was a concept long since abandoned. In the recently constructed plants, the walls of the fire chamber had become substantially a complex of tubes. The water entered at the bottom, was subjected to intense heat, and emerged at the top of the chamber as steam, although not quite yet ready for use.

In this newest unit at Monroe, the last vestige of the older system, the so called drum, disappeared. The essential fact is that the leap from 2,500 psi to 3,500 psi at 1,000 degrees carried the whole operation to the pressure area called supercritical, in which the distinction between steam and water disappears, and with it the problem of separation and a good deal of the attendant equipment. The saving in fuel is small but, in an operation of such size, it became important.

The steam generators for the four units were all ordered from

Monroe Power Plant under construction (1970). In the foreground is a unit train, and the coal pile.

The peaking units were small, special purpose machines. Nine such installations (1966-68) had a total capacity of 325 OOO kw.

Babcock and Wilcox, with whom the Company had had frequent business relations. The turbine generators No. 1 and No. 4 were ordered from General Electric, and machines No. 2 and No. 3 were manufactured by Westinghouse. Bids had been sought from both Britain and Switzerland, for Units No. 3 and No. 4, but the contracts went to the American companies already involved at the site.

The turbines, although made by two different companies, were of similar general appearance and capability; they are basically extrapolations of the machines most recently purchased. Three sets of turbines, for high, intermediate, and low pressure, are mounted on a single shaft. The high pressure unit will receive steam at supercritical pressure and at or a little above 1,000 degrees. The discharged steam will move to the reheat unit and return to 1,000 degrees before it enters the intermediate and then the low pressure units.

These turbines are mounted tandem on a single shaft, and at the other end of the shaft from the turbines, the generator unit is mounted. The whole complex rests on a massive concrete base, 41 feet across. There is a slight difference in over-all length due to the different manufacturers, but the shafts are roughly 175 feet long. The entire unit, including the turbines, generators, and support, must be built to tolerances more characteristic of watchmaking than of such massive structures.

One feature certain to call attention to the Monroe project is the pair of stacks which will serve the four units. Like the plants themselves, both size and innovation add to their interest. The stacks are poured concrete, self supporting during construction, tapering pleasantly with increasing height.

Electrostatic elements will catch and remove practically all ash and dust, there will be no smoke, and the gas residue of combustion will be dissipated at a height almost one sixth of a mile above Lake Erie. The stacks will stand on a base 75 feet in diameter, taper to a diameter of 35 feet, and rise to the equivalent of a 60-story building.

The earliest years of the electrical industry were troubled by the illusion that electricity could some way be a no-cost by-product of some other activity. Men tied it to salt making, to lumbering and to central heating service; by far the most long lived of these dreams related to conservation and the exploitation of no-cost water power. One writer has referred to this early wide spread, as "hydro-mania."

Rather early it became clear that there was no assured economy even in hydro-electricity, except where subsidy and tax advantage were used. Where the year-round supply of water was adequate, and where location did not inflict too heavy a handicap, hydroelectricity was economical; the saving as against a modern thermal plant, however, was less than enthusiasm first indicated. The critical limiting fact was that hydro-power could not be indefinitely expanded to keep up with load growth.

There were, however, some very real advantages in a system which contained both hydro-power and thermal powered plants. Ontario Hydro-electric had long since gone to this combined power production, and Edison was close enough to that system to share in the advantage. Start-up for a thermal plant can take a number of hours; it is expensive and inflexible. The hydro-plant is almost instantly operative when the water begins to flow. Even a small or limited supply of water, properly conserved for use at peak hours, can be of service in a system fundamentally dependent on a basic thermal production.

Abroad, and at a few places in the United States, an ingenious modification of this use of hydro-power had been developed which could give marked advantages to the Michigan area. The word "pumped-storage" began to appear in the discussion of Company problems well before the 1962 pooling agreement, but became a reality in planning soon thereafter.

Engineers of the Consumers Company had located a desirable site for such a unit and, after the pooling agreement, Edison engineers agreed with the judgment. South of Ludington, on Lake Michigan, the shoreline bluff has an elevation of roughly 200 feet. The site selected had no lake or flowing streams. It was proposed to put a tank on top of the hill, and let the flow of the water from the tank generate hydro-electricity in a power house on the shore of Lake Michigan. Then electricity would pump the tank full again.

The engineers agreed that the pumped storage unit will demand four kilowatt hours of input for every three kilowatt hours returned. The important thing is that the four kwhr will be cheap ones, at the hours when they can be spared, and the three kwhr output can be timed to the hours when demand is greatest, and in amounts promptly and efficiently adjusted to the demand.

The pumped storage system is sometimes explained by the analogy of the storage battery. From the industry's point of view, the advantage is the flexibility and promptness with which it can respond. The Ludington project, huge though it is, is basically a "peaking" unit which promises, as those expensive short hour operations are judged, to be economical both in installation costs and in operation.

The structure itself can be described simply. On the bluff above the lake, a six mile dike will surround an artificial lake. This is a huge earth moving operation, for the dike will be 400 feet thick at the base, and as much as 170 feet high. This dike conforms to the contour of the land to bring the top to a level 370 feet above the level of Lake Michigan. The inner surface of the dike will be paved with a special asphalt material.

From the artificial tank to the Lake Michigan level, water will flow through six steel tubes 1,100 feet long. These tubes will taper from 28 feet to 24 feet as they approach the turbines. In the power house, the six turbines will have a capacity of 312,000 kw each, to make available any amount of power up to 1.87 million kw on demand of the two systems. The full flow of water will be at the rate of 33 million gallons per minute.

When electricity is available in surplus, the turbines will become pumps and the generators become motors, and water can be pumped up these same penstocks to the tank lake above. It is expected that the daily outflow of water will exceed the input throughout the week, with the period of low system demand over the weekend allowing full restoration to stored capacity.

Both American Electric Power and Commonwealth Edison of Chicago were interested in the plans when they were announced in 1967. The construction of this facility—which is in Consumers' territory—was the responsibility of that company; Edison will share in the capital costs to the extent of 49 per cent. The total cost of the project, $186 million at first estimate, appeared modest in view of the services expected. The project was of a size which would not have been justified in either company alone.[15]

The final agreement between the companies was delayed to 1969, but that was on detail; designing and plans had been under development for several years and the actual work had already begun. The completion date was announced as 1973. When the project was announced by the companies, the *New York Times* remarked that, when completed, it would be the largest of its type in the world. At the moment, it reported, the largest such facility in operation was the 800,000 kw plant of Philadelphia Electric at Muddy Run.[16]

Ludington Pump Storage is a creative contribution to the power problem of the pool. It exploits the basic fact on which the industry had insisted so long—the cost of electricity is not a simple and fixed absolute.

While the Monroe plant was in the early stages of construction and the Ludington project was still at the discussion stage, the second plant on the Fermi location was announced.

The history of the first Fermi plant is a separate story. (See Chapter Sixteen) It was the joint activity of a score or more utilities and business firms, designed to bring to acceptability the technology and the economy of the breeder type reactor. The breeder reactor would operate at high temperatures, would be economical in fuel, and would possess other advantages but, by Company estimate, the commercial solution of its problems in reliability and economy were still a decade away.

In the meantime, the industry elsewhere had led in the development of nuclear power in a much simpler form fundamentally on naval reactor technology. Detroit Edison had supported some of the research done, making its contribution with other companies and industries.

The decision in 1965 and 1966 to use coal for the huge Monroe plant had been basically economic, the weighing of the price differential between coal and nuclear fuel. Between 1965 and 1968, however, a rather dramatic alteration in delivered coal prices, together with estimates of future trends, tipped the scale, and the decision this time was nuclear energy, with a reactor of tested commercial capability.[17]

This new plant was located on grounds a few hundred yards from Fermi No. 1, on the shores of Lake Erie. Space was also reserved on the same site for a full scale breeder reactor plant when technology permitted. An analysis, in a press release, noted that the two nuclear plants would, in the long run, be complementary in their fuel characteristics.[18]

One important characteristic of the nuclear plants so far was the rising efficiency and economy with increased size. The same characteristic had marked coal burning plants at an earlier time, but there appeared to be a limit to such increases in efficiency with coal. In the atomic energy plants, at least up to the time of the planning of Fermi No. 2, no such limitation had been reached.

The atomic reactor was ordered August 15, 1968 from General Electric, and was of the water cooled, slow neutron type. Steam production was calculated at 14 million pounds per hour. The schedule called for operation by the end of 1974. Before the end of 1968, details of the project were being prepared for submission to the United States Atomic Energy Commission for examination and approval of the construction permit. The Commission would exercise close control over each step in the program.

For the third time, the Company ordered a turbo-generator from abroad. The order went to the English Electric Corporation of Stafford, England, by way of a special subsidiary of the same name set up by the British firm in New York. The Company employed a group of Belgian experts—some of whom had worked on loan from their own companies on Fermi No. 1—to supervise the manufacture.

The turbine generator ordered was an enormous piece of machinery. When design became fixed, the Company referred to its capacity as 1,150,000 kw. Less than ten years before, the largest machine in a Company plant was 325,000 kw, installed in 1961 and at that time among the largest in the industry. At the opening of World War II, the total capability of Detroit Edison was hardly more than that of this one machine.

CHAPTER FIFTEEN

FUELS AND THE ATTENDANT PROBLEMS

Coal and the Elevation of Energy / The Pattern of Transportation and Use / The Pipe-line Tariff and the Unit Trains / Monroe and its Fuel System / Pollution; The Control of Smoke and Grime / Pollution; Sulphur and its Compounds / The Costs and Extensions of the Programs.

THE ELECTRICAL INDUSTRY DOES NOT CREATE THE ENERGY IT SELLS; ITS entire function is to upgrade energy from a lower to a more usable form, and to make that energy available at a point of consumer demand.

The conversion of energy to this new form must be done at a cost which the industry will find acceptable, and without waste of any important natural resource. In this matter, as is so often true, good business practice is also socially desirable practice. This idea must be emphasized especially in any discussion of fuels and pollution.

In the United States, the electric utility industry consumes more than half of the coal mined, and the proportion is rising. No discussion of the industry's performance in this field can be complete without consideration of alternative methods by which so much coal would become usable energy, or without a comparison of the consequences of alternative use on either conservation of resources or the quality of the environment.

By an easy rule of thumb, the total expenditures of the Company are divided into five roughly equal parts: fuel, taxes, wages, supplies and material (including future retirement of property), and interest and dividends (including earnings retained in the Company). In 1961, the cost of fuel had been exceeded by both taxes and payroll; in 1968, for the first time, coal and its equated fuels was the largest of these five categories.

In 1968, Detroit Edison was number four among the privately owned utility systems in the total amount of coal used. Of all of the coal produced in the United States, Detroit Edison purchased one ton out of forty-three mined, and the total purchase price approached one hundred million dollars per year.

Coal is the primary fuel in The Detroit Edison Company; in special cases, oil and gas have supplemented or supplanted coal, but the total amount is not large. Some electricity produced by hydro-power enters the system over the Ontario interconnection, but this amount is also small and is a decreasing percentage of the total. As yet, nuclear fuel is a hope rather than a current reality.

Up to and including the decade of the fifties, the methods of handling coal had become standardized, in part because of the relatively small amount involved. The small, unmechanized mines of a low capacity were characteristic suppliers; the Company estimated that it bought from one hundred and fifty different mines.

The method of delivery for all plants was substantially the same. Coal was shipped by rail to Toledo, where it was dumped into self-unloading barges or boats for lake shipment. Each of the four old plants had docks for receiving coal, and of course, the newer plants, River Rouge and St. Clair, when they were completed in the fifties, had automated dock facilities.

Depending on the destination, the use of the rail-lake delivery saved from fifty to eighty-five cents per ton over the then existing single car, all rail tariff. Railroad facilities had been essential when the plants were under construction, and in normal course equipment was installed which would allow the handling of coal by rail also. St. Clair, the only plant without facilities for receiving coal by rail, was an exception and thus it was totally dependent on water shipment.[1]

The coal ships and barges shuttled between the Edison plants and the Toledo coal dock through the entire summer and into the autumn with fuel for current usage and for the winter stockpile. After the closing of navigation, there was a continuing flow of coal by rail at a much reduced volume and the mountains of reserve fuel about the plants were gradually reduced to the minimum protective level.

Three plants still contained generating equipment installed before World War II, some of it long before. The boilers for these old plants were stoker fed, and they required a premium quality of coal which was also more expensive. The critical characteristic required was that it had a high fusion point, and that at normal operating temperatures it did not form massive clinkers. This coal was brought from Virginia, West Virginia and Kentucky and was referred to as southern coal.

This old equipment at Marysville, Delray, and Conners Creek was expensive to operate, both in man power required and in coal consumed. The units themselves were small, by modern standards; the stoker fired boilers served eighteen old turbines. In the days of reduced demand, such as had characterized the half decade before 1962, these machines were frequently idle, or were operating as reserve only.

Trenton Channel, the pride of the Company in the twenties, was the exception; it utilized a new and quite different method of firing. Coal was run through a pulverizing mill; the air born coal was then forced into the furnace on a blast of super hot air, and literally burned in suspension. The great advantage of this method was that it could use efficiently the less expensive Ohio coals, and that the explosive heat allowed operation at a higher temperature. The great disadvantage was that too much ash went up the stack. Pending the solution to this difficult problem, the next building (at Delray 1929) reverted to the older type, the stoker.

Following World War II, the technological advantage of the use of pulverized coal became so marked, and the progress made in the control of flying ash so promising, that the new plants and additions to the old ones all used this method. By 1961, the stoker fired units were rated at only 840,000 kw of the 3,836,000 kw capacity of the Company. The new units added in 1968 and 1969, of course, were of this type. Including central heating plants, which were also stoker, the Company when it was utilizing every facility in production was consuming roughly thirty percent southern coal, and seventy percent Ohio type coal.

In spite of generally inflationary prices after World War II, the coal industry did a remarkably good job of controlling price increases. With the loss of coal buyers to gas and oil, especially in the railroads and homes, the industry from about 1948 to 1961 was capable of over production, and felt intense competition both from within and without. The intelligent action of the union in allowing full mechanization greatly increased productivity per man and became the basis of a wage adjustment which was mutually satisfactory. The cost of coal, measured either by the ton or by Btu content, remained remarkably stable for a decade following the postwar price adjustment of 1948.

For the Company, the cost of the coal at the mine mouth was only a part of the cost of fuel, and sometimes it had been the lesser part. On the lakes and river, the presence of competing ship and barge lines allowed increased efficiency in operation to show itself in reduced freight rates. Larger size boats also reduced costs.

On the railroads, circumstances conspired to maintain both the freight pattern and the rates. Among the major railroads, the Detroit-Toledo area was served by the C & O, B & O, the New York Central, and the Pennsylvania. Since coal shipments originated at the mines, for a long haul, the roads had small interest in a joint traffic with other lines which would introduce coal produced elsewhere. For practical purposes, their control of the joint rates confined Edison's purchases to the two coal regions, in spite of the so-called mid-western fields in Indiana and Illinois.

The rates continued to be those which reflected the low volume prewar practices. The Company paid the price which would have been paid by a dealer who ordered one or several carloads at a time. The rather static regulatory policy of the federal authorities supported the efforts of the railroad association to maintain the highly satisfactory pricing system. Between 1954 and 1958, the rates for shipment had gone up rather sharply. The increase in February, 1958, came at an inopportune moment, for with the slowing of the economy in the Detroit area the Company was earnestly searching for means of saving money.[2]

Utilities and other major consumers of coal, especially in the Lake Michigan area, had been exploring the possibility of the use of the Illinois barge canal for access to the western fields. The Chicago and Eastern Illinois Railroad agreed to provide mine-to-barge service; at Chicago the coal could be shifted from canal barge to lake vessels. Though the distance was five times as great, the overall cost proved to be less, and Detroit had achieved a competitive situation in both supply and transportation of coal.

The eastern railroads made the effort to deny Detroit the advantage of this competitive arrangement by establishing a line on the lake beyond which this coal could not go. It was an obvious conspiracy and Robert Lundgren from Fuel Supply and Harvey Fischer, Michigan counsel, held a discreet conversation on the subject in the offices of the Department of Justice in Washington, and allowed the rumor about the matter to circulate. No further action was necessary, for overnight the restriction was lifted. The eastern roads then moved to meet the new price competition. The new rate reduced the cost of coal at St. Clair by thirty-seven cents per ton and at Detroit by fifty cents per ton.

In October, 1960, the presidents of the C & O, the Norfolk and Western, and the Pennsylvania all asked for meetings with Cisler.[3] The immediate emergency was the permission granted by the Interstate Commerce Commission to further increase rates; the roads proposed a negotiated special rate for a larger tonnage. Beyond the immediate, however, there was a more significant development by which the railroads might well be deeply affected.

For the first time, there was an alternative available to complete dependence on the railroad systems. There had recently been some speculation about, and some experience with, a plan which would locate the power plant at the mine mouth and transport power rather than coal. At the moment, however, this was a dubious proposal for Detroit Edison.

There was even more interest in the newer proposal by the Pittsburgh Consolidation Coal Company to substitute a pipe line for the railroad. At first glance, the method seemed to be only an extension of the current practice in pulverized coal. It was proposed that the coal be pulverized at the mine, mixed with water to become what was called "slurry" and pumped to the power house.

A special study by the Company, with the cooperation of one of the major suppliers of coal, concerned only the Trenton Channel and River Rouge plants. The conclusion was that on the four million tons required annually the saving would be estimated at $1.25 per ton. This study was made available to the railroad officials, who quickly lost their complacency.

Negotiations were extensive, and in time came to include all seven of the railroads involved in this part of the coal region. The contract was the first which took realistic consideration of the reduced costs which massed shipment made possible. The established price was cut 37½ cents for the first million tons shipped and 50 cents thereafter. For so fundamental a concession, the Company had to respond. The contract became effective in October, 1961, and bound both parties for a period of ten years, during which the Company agreed not to introduce the pipe line project into any of its plants, in return for substantially reduced rates. This contract became a basic document in defining relations with the railroads on which future negotiations would rest. In the Company and perhaps also in the railroads, this agreement was referred to as the pipe line tariff.[4]

On November 26, 1962, President Stuart Saunders of the Norfolk and Western, addressed the EEI convention. His road was one of the coal handlers, so he had a personal concern, but he also spoke on behalf of a troubled industry.[5] Coal was twenty-five percent of the total freight tonnage of American railroads, and for the first time coal shippers had alternatives. The railroads must institute real economies to offset the price cuts. His appeal to the electric industry was for time to achieve system integration, which alone would allow the consolidation of railroads to become effective, and especially to allow the full development of the new idea of the unitized or integrated train.

A group of thirty-five railroads had already done some work of an experimental nature, with most encouraging results. In the summer of 1962, both the New York Central and the Pennsylvania announced the establishment of a new service, including new lower rates. In each case, the train would run, without yarding or switching and at substantially passenger train schedule, between a designated source for coal and a single destination. The rate reduction, as the Pennsylvania Railroad announced it, was to the order of twenty-five percent.

In May, 1962, the executive vice president of the New York Central, Wayne Hoffman, met to accelerate a study of the possible application of this new system to the Edison area. Edison had a good price position, for the contract (the so-called pipe line contract) and the availability of water transportation for at least part of the way gave that Company an advantage. The Baltimore and Ohio, with the cooperation of the Company and the coal supplier, began effective experimentation looking toward a price negotiation.[6]

The railroads, with their involved interregional tariffs, were slow in achieving agreements. Not until October and November, 1963, could they come up with a proposal which became the base for negotiations. It became clear, even in this preliminary discussion, that the savings in freight would be to the order of two million dollars per year. The acceptance of the terms in February, 1964, was on specific condition that the seven railroads involved understood that acceptance did not prohibit further negotiation by the Company on either price or conditions with any one of them. The contract itself, including even its duration, was subsequently modified by agreement.[7] By the end of 1964, the unit train was a reality, running between the mine area and the lake shipping point.

It is possible to identify several changes in Company practice which the unit train produced. The achievement of economy depended on the rapidity with which the round trip could be made between the mines and the docks. The cars must be loaded promptly, at one or two sources. Since the many small mines lacked the facilities and the volume required, the number of possible suppliers was reduced in number.

Flexibility was further reduced at the receiving end of the supply line. In the lake season of 1965, the Company received 1,120 vessel shipments, a new high in Company usage. In that year, the Company purchased just over ten million tons of coal, of which all but ten percent reached the plants by the combined rail-lake routes.[8] The ideal coordination of boat and train schedules to avoid a second handling of the coal was hard to obtain.

With the dependence on the lake-rail route, the huge stock pile of coal at each plant was a costly requirement imposed by the closing of the navigation season. When the Company became the sole or the primary customer of a big deep mine, the closing of navigation also closed the mine, and became a real hardship both for the mine operators and the miners. An all year operation would be advantageous for both, if it became possible.

The opportunity to negotiate such an operation came when several new generation units in Edison made necessary provision for additional coal. The first all rail unit train to the Detroit area, in the winter of 1964-65, was under this new contract. The saving by the use of the new type train was seventy-one cents per ton, as contrasted with the rates for single car load lots. The surprising comparison in rates was with the land-lake route rate, even with the prevailing pipe line tariff. The new unit trains, once loaded and rolling, could extend their run beyond Toledo to Detroit at minimum cost, and without the reloading operation in Toledo.

In spite of the inevitable difficulties of a new and highly integrated operation, the results justified the efforts. In a time when prices were generally rising, it was possible to report to the board in February, 1965, that the price of coal, delivered, was the lowest it had been since 1955.

The requirement that the coal be unloaded in a specified number of hours emphasized what was basically the new element in the whole problem of coal transportation—the element of time as against mere distance. For the railroad itself, and for the Company, speed in handling both in transit and at the ends of the lines increased the capacity of a given equipment and thus reduced the overall cost. It was clear that here, in the perfection of the handling operations, would come the next series of technological improvements.

In the two or three years which followed this first experimental run, a series of changes were tested. The necessity for complete command of the requisite number of cars forced dependence on shipper-owned or receiver-owned equipment, rather than the standard railroad coal car. This led to the development of the rotary coupling and other special equipment and methods.

For the railroad, both the fact that the cars were provided without investment responsibility and that they allowed a more rapid turn around were important in the calculation of costs. A further and highly significant point in the development of the contract was the high volume of freight which was guaranteed and the long term which the contract covered. The railroads became sharply competitive for the new and growing business.

One contract tied the Georgetown mine of Consolidation Coal Company to Trenton Channel, where machines 7 and 8 were in operation and No. 9 was to be activated in 1968. Two trains on a three day round trip schedule brought eighty-two leased cars, each carrying one hundred tons of coal, from a single mine to a single plant. A similar contract tied three cooperating mines to the important River Rouge plant, also with two trains on a three day round trip from mines to the single plant. Utimately, as will be indicated, a similar arrangement on an even more elaborate scale would serve Monroe.

There were problems in this kind of arrangement. The new equipment was untested, and at first troublesome. Dependency on a single mine could pose serious delays if there were labor troubles or an accident there. A rail interruption could disturb transportation arrangements. At both Trenton Channel and River Rouge plants, equipment remained in place with which to handle coal by water if that should become necessary. At St. Clair, with the addition of unit No. 6, a unit train facility was added as a precaution against interruption of delivery by water.

This had been a remarkable commercial achievement. The availability first of alternative routes, and then of alternative methods of transportation had opened the way to negotiation with the roads on a new basis. Out of the negotiation had come joint efforts which reduced railroad costs, and allowed further adjustments in tariff.[9]

At the end of 1968, with these rate reductions and economies in effect, Robert W. Lundgren, then manager of Purchases, summed up the effect of the adjustments for the Company. The 1968 tonnage of coal, carried at the rates then current, cost $16.5 million less in freight charges than would have been paid at transportation rates of ten years before.

The year 1965 saw the lowest total cost for fuel, measured in Btu content, since the postwar price adjustment in 1946. By the end of the year 1969, the report was less optimistic. In that year, the Company had burned in coal (or its equivalent in gas or oil) 14.5 million tons, an increase in the year of 10.9 percent. The significant item was the increase in fuel cost per Btu, which had risen annually since 1965. The rising costs in coal at the mine, and the burning of more of the more expensive fuels in places where coal was undesirable, combined in roughly equal degree to raise the costs sharply. Lundgren pointed out that the entire saving from the achievement of extraordinary transportation efficiency had been offset by these rising costs. The cost of fuel, per million Btu, was close to the highest in Company history, and the trend seemed to be continuing upward.

The development of the plant at Monroe brought sharply to focus the problems with which the Company had been struggling for half a decade. For the first time, in any significant sense, there was not just an alternate transportation route, but there was an alternate fuel; both the coal producers and the transportation agencies were involved in the competition.

By the decade of the sixties, the use of atomic energy was well past the experimental stages, and had been subjected to engineering studies. Inside the Company, with its own people, and then through the use of Constantine W. Bary, economic consultant to utilities, the costs of construction and operation of a large atomic plant was compared to the equivalent fossil fuel in production from coal firing.

The proposal here was not the use of the fast breeder reactor, which the Company and associates had been struggling to develop. In time, this could prove both practical and very economical, but there were still too many unsolved problems, and the system could not be accepted as commercial. There were, however, several types of conventional nuclear reactors which had been proved competitive with coal in other utility systems. As a product of premature enthusiasm, very attractive estimates for both costs in construction and in operation were available.

It was possible to establish within narrow limits the anticipated fuel costs per million Btu with atomic energy, and from that to set a target which the coal suppliers must reach if they were to be competitive. If coal were to supply the fuel for this projected plant, then both the producers and the transporters would be put to it to meet the challenge.

Negotiations began in the summer of 1966 with the representatives of Consolidation Coal, the New York Central, and the Pennsylvania Railroad. The target set by the Company was a price of twenty cents per million Btu, which at first seemed to the others impossible of attainment. By November, 1966, however, Bary was advising that the newer offers for the sale and transportation of coal substantially eliminated the cost differential, and that under the circumstances the familiar coal burning plant should be

Donald F. Kigar President 1964-67.

Edwin O. George President 1967-70.

William G. Meese, President 1970—.

planned.[10] Substantial agreement had been reached, but there were many details to be worked out before the final contract would be signed.

Involved in the announcement (December 18, 1968) were L. F. McCullum, chairman of Continental Oil, of which the coal company was a subsidiary, George H. Love, chairman of Consolidation Coal, Stuart T. Saunders, chairman of the Pennsylvania Railroad, Alfred Perlman, president of the New York Central, and Walker L. Cisler. Each man spoke for a prestigious company, which he was committing to the service of the new facility at Monroe.[11]

The problem of Monroe was vastly difficult, if only because of scale. Each unit would demand two million tons per year, and the original two units planned grew to three and then four. To supply coal in such quantities from a single or even several sources demanded an organization in the mining area of the sort not available.

Consolidation Coal had coal reserves in abundance, but they were not developed, or indeed available to railroad transportation, and that company either lacked the resources or was unwilling to finance the operation. Extended negotiations involved discussion of a jointly owned subsidiary company, of a purchase-lease agreement, and of other innovations. The end agreement was a simple loan from Edison on favorable terms, with principal and interest payable in coal shipped.[12]

The trains themselves turned out to be technological innovations aimed at that problem of time. At the plant, the unit train swings in on a loop, to position cars one after the other at the dump. A rotating coupling between cars allows the car to be tipped, emptied and righted again in a fraction of a minute, and without being disconnected from the train. For all practical purposes, the train hardly stops either at the mine for loading or at the plant for unloading. A seventy-two hour schedule, for a seven hundred mile round trip, leaves no surplus time.

The integration of industrial processes is, of course, the essence of the so-called factory method. This was much more complex, for this was articulation on a new scope. According to Walker Cisler, "Coal production, transportation, electric energy production and financing as planned . . . constitute an entire economic system in itself."[13]

The word pollution has been most loosely used, with resulting confusion. It has been used to describe a little harmless white dust which may distress the careful housewife. It has been extended all the way to include the possibilities of an accident, with untold lethal results. Within recent times, the conception has been extended beyond air and water, to include thermal change, noise, landscape defacement, and indeed the entire range of environment protection and control.

For industry everywhere, there was danger that public authorities, in their eagerness to do something about pollution, would not recognize the tentative character of research and might order into effect methods not yet tested. Premature legislation of administrative ruling could, if it imposed rigid standards, inhibit further technological improvement.

With all of this, both the Company and the industry were concerned; they were prepared to spend millions of dollars, but they needed the assurance that the need justified the cost, and that the technology employed would accomplish the intended purpose. No one would dare equate human safety to money in any fashion, but sometimes the complaints were of a mere annoyance, and the costs of its suppression disproportionally high.

Over the years, the Company had been concerned chiefly with the most obvious of the irritations. The first of these is smoke, and the second is the emission of solids from the stack in the form of a fine dust or ash. When the pure carbon in coal is perfectly burned, under ideal conditions there is small residue and no pollution. The problem to which so much angry attention has been given comes from the fact that such perfection seldom occurs outside the laboratory.

To the Company, the tests which used color of the stack emissions as a test of the extent of pollution were inadequate. They were relatively easy to enforce, however, and sometimes governmental agencies, especially on the local level, tried to use the visual scale. The protests of the Company were not against the regulation, but against the measurement devices which were used in the regulation.

A huge black plume of smoke from a stack is the highly visible mark of bad burning; volatile materials are being wasted rather than consumed, with attendant financial loss. The owner of the stack has a financial interest in correcting the situation. When this happened in the Company, barring employe error, the cause was usually ancient equipment which was being pushed beyond the proper limits to meet the increased load.

The interests of the Company and of the neighborhood are identical in this matter. Under normal circumstances the more modern plants, including Rouge and St. Clair raised fewer problems, and the technical competence in the control of this nuisance was more complete. The smoke problem was confined to the older machines. When and as they could be retired, Edison would have solved the problems of smoke.

Something of the same could be said about the particulates in the stack emission, though there the problem is a little more complex. Since the building of the Trenton Channel plant, the men of the Company there have been tormented with the question of fly ash. This plant, built in the twenties, was the first in which powdered coal was burned in suspension. Without some kind of artificial control, the flame would have drawn one third of the ash up the stack and scattered it over the neighborhood.

Trenton Channel was equipped with the so-called Cottrell electrostatic dust collector, developed in part with the cooperation of the Company and, so far as is known, first used in a power plant at Trenton Channel. It took time to perfect and adjust this mechanism, but gradually its performance improved, and the problem of ash and other particles came close to elimination with this kind of equipment.

After World War II, the burning of pulverized fuel became standard, and in the units installed in the fifties at St. Clair and River Rouge, the removal of particles reached an efficiency of 98 percent. The combined electrostatic and mechanical dust collectors for the new units, 1968-1969 were projected for 99.6 percent efficiency. For practical purposes both smoke and the particulate nuisance had been controlled in the new equipment.

Unhappily, the old plants still survived. The old stoker fired boilers could no longer measure up to the standards being set by modern machines in the matter of emission of solids, and they were especially liable to smoke violation. These plants contributed coal dust blown from their mountains of stored fuel, and solids blown from the stack, and the nearby regions were likely to blame the plants for the total accumulated grime of the city.

Through the decade of the sixties, the Company had twenty-one machines on the line which predated World War II, fifteen of them with old type stoker fired boilers. The equipment which so successfully curbed the ash problem in the new plants could not be adapted or installed in the older ones. The cost of such installation would have been prohibitive, and the discharge of smoke would not have been diminished.

These older plants or sections of plants were most obviously in trouble. Abandonment of at least some of these machines, or at least their retirement to reserve, was made impossible by the racing increase in demand for power. The Company faced the necessity of using them, and that meant doing something about them, where and as it could. When possible they were given greatly reduced loads, in the hope that at that level they would perform within acceptable limits. At great expense, the Company used its oil and gas fired peaking machines to supplement production at those hours which otherwise would have put pressure upon the reserve units.

Some combination of fuels seemed called for, and preliminary tests were satisfactory. At Conners Creek, which was the most critical point, because of its location, the experimental use of oil and gas burned over the bed of coal on the stoker greatly reduced the emission of particles from the stack and plans were made to extend the system to the entire plant. The equipment installed allowed the full use of gas or oil as substitute fuel when availability and/or price permitted.

The only method of smoke control at Delray, was the elimination of coal firing. Gas was chosen as the substitute. The contract with the gas company called for an "interruptible" supply of fuel, in order to qualify for a reasonably priced industrial rate. The Company installed oil burning equipment also, as a precaution against temporary interruption of the gas supply.

The problems associated with pollution of the air are old in American industry, and the Company shared in the concern. In 1953, W. A. Carter, of the production department of the Company, had been elected to the board of directors of the Air Pollution and Smoke Prevention Association at its 45th annual meeting in Cleveland.[14]

The city of Detroit had an ordinance on the subject, and a Bureau of Air Pollution Control, but the enforcement problem was difficult. In 1962, the Bureau was still attempting to use the controversial "Opacity Test," which was hardly more sophisticated than a color chart. The legal department of the Company worked with the Greater Detroit Board of Commerce and its air pollution committee in an effort to achieve a more realistic ordinance.[15] In November, 1965, the proposed control legislation for Wayne County posed the same problem of acceptable standards. In October, 1966, the Board of Health for the State of Michigan issued its rules on air pollution. Some individual cities of the Edison area set up their own limitations.

Almost all of these regulations had to do with smoke, color and emission. There was danger that contradictory instructions from conflicting authority could develop, but the Company attempted to stay ahead of the legal minimums. Every plant built after 1951 was designed to meet standards which still maintained as late as 1966, though in operation, especially under maximum production, those standards were not always achieved in the older units.[16]

The program for cleaner air brought unexpected benefits to the Company. Detroit had its share of those industries which burned coal. More rigorous enforcement of pollution regulations forced suspension of operations, or the installation of new equipment and many of these industries turned to electricity.

Far more difficult to deal with, and potentially much more dangerous, were the gaseous materials which came chiefly from impurities in coal. Of these, sulphur is the prime offender in the so-called northern type coal most commonly used in the Company. In the form of sulphur dioxide it was both the most common and potentially the most dangerous of the gases discharged from the stack.

Under favorable circumstances, the gas would be discharged harmlessly, with the wind about the stack scattering and diluting it. Sometimes, however, climatic conditions and/or the presence in the atmosphere of other significant materials brought about a chemical reaction, and a shower of sulphuric acid, in mild solution—would fall on neighboring homes and yards.

The high concentration of sulphur dioxide under certain meteorological conditions led to an engineering study of the possibility of increasing the height of the structures already completed. Air tunnel experimentation confirmed the Company's experience that the increased height of the plant's stack contributed to the reduction of the concentration at ground level. As new stacks were built, both in old and new plants the tendency was to reach for greater and greater elevation. But this was not the ultimate solution to a problem which, with the increasing size of plants, could only become difficult.

In November, 1967, Cisler and a small group of other executives from American power industries visited Britain for a conference with authorities there, in which the problem of sulphur control was foremost. There was a rumor that the British had a successful removal technique, but the report on this seemed to show that it was neither fully effective nor economically justifiable. Cisler's correspondence after the conference confirmed this. What the British had undertaken was an extensive observation of sulphur concentrations at ground level, and at observation posts at differing distances and directions—a reporting device long in Company use. The high stack still seemed the only practical and available control, although alternative solutions were being explored by the industry.

The Company's own research was supplemented by a cooperative effort with the coal supplier, and by financial support to joint efforts through an association. The introduction of atomic fuels would promptly clear up this problem but this was not at the moment a politically practicable alternative. Nor was the use of the coals containing little or no sulphur the answer, for there was literally not enough to go around.[17]

Sulphur itself is a marketable commodity, if it could be extracted either from the coal or from the flue gases. Cleaning the coal before burning required very fine powdering as a preliminary; it was neither effective nor economically promising. Despite claims by the venders of the equipment, the Company is not yet persuaded that the methods of flue gas removal would adequately control the sulphur, or that the cost was tolerable even with the sulphur as a byproduct.[18]

In the absence of alternatives, the stacks continued to rise.[19] At Monroe, where the new technology was most completely exploited, the stacks rose to 800 feet, just a little less than one sixth of a mile, higher than any building in the state of Michigan and almost fifty percent taller than the Penobscot Building in downtown Detroit.

Pending the beginning of operations in the Monroe area, where both the Monroe coal burning plant and the Fermi No. 2 nuclear plant were under construction, the Company undertook an extensive study of both air and water as they existed prior to the beginning of operations. In part, this was to help substantiate the fact that the Company has sometimes been falsely accused of pollution. More importantly, it would enable the Company to measure its performance in a field in which it has significant interest, and on which it has spent so much money.

The first Federal legislation on air pollution was enacted in 1955, and was chiefly consultive and advisory. The Clear Air Act of December, 1963, was amended and much extended in 1967. This enactment, popularly referred to as the Air Quality Act required the Department of Health, Education, and Welfare to establish control regions for various pollutants. It then became the responsibility of the state to set standards and implementation procedures. The original criteria which HEW issued were for sulphur oxide and particulates, the problems on which the Company had been working.

For the Company, there could be a marked advantage in a single set of criteria, and a uniform regulatory policy; the standards imposed, even if strict, would be explicit. Criticism of the Act, in its early operation, has been directed chiefly at the diversity in standards and enforcement procedures as between the states and regions, and implies the desirability of a single national standard on pollution control.

The attention to air pollution was obvious, but in Michigan the popular concern with pollution of lakes and rivers was just as vocally expressed. In the main, Detroit Edison was not concerned. Each power plant drew water from the river, used it to cool the condensers, and returned it to the river a little cleaner and a little warmer for the use.

That shift of temperature, for obvious reasons, was called thermal pollution, to take its place along side of particulates and gases. Recent legislation required the company, and other industries of the state, to have approval for the use of lake water. This was not a problem at Harbor Beach. Permission for the same right at Monroe was different. The difference was made obvious by the largeness of the plant and the smallness of the lake, which was also very shallow and already seriously damaged by more conventional pollution.

Thermal pollution is a problem about which almost nothing is known, although conjectures and prophesies regarding it have been many and dire. Given the climate of the great lakes, the problem is unimportant in the extended winter months, although some enthusiasts will not grant even that. Beyond that fact, no one had studied the impact of thermal change in such proportions on plant and animal life, and that was the concern of conservationists.

The Company had begun its first studies in the ecology of the Lake Erie region in 1954, in anticipation of the first work on Fermi No. 1. In its later application for access to this water, Edison proposed the establishment of a new study program, in association with teams from the state's universities, for at least a period of five years. The license, when the Company received it from the State Water Resources Commission (1969), made that research on Company subsidy a condition of the grant.[20]

This whole new group of related activities called for coordination. In January, 1968, a new position was created on the staff of John W. Drummond, vice president and manager of operations; Wayne L. Wingert was named Environmental Improvement Engineer. For the moment, his attention was demanded by these pressing problems of pollution of air and water, but specifically his area of concern extended to the whole problem of environment. This would include such matters as land use and development, the location and appearance of tower lines, and kindred subjects; such concern was necessary, in the words of Walker Cisler, if we are "to more fully discharge our responsibilities . . ."

The cost of such a program for pollution control, or even a summary of past expenditures, is almost impossible to establish. The correction of the performance of the older plants, including the industrial plants purchased in the decade, had been very expensive.[21] Ed George made the educated guess in 1968 that the expense of air quality control since 1924 had been $30 million, and that the current program would cost $20 million more. Including expenditures forward to 1973, Walker Cisler estimated the total cost at $63 million. There has been no federal subsidy in this operation.[22] Concern with the whole complicated problem led management to consult the National Sanitation Foundation, a distinguished organization in the field of environmental research. At the Company's request, this organization named a panel of five experts to review the control endeavors of the Company generally, but especially in relation to the new Monroe plant, and the proposed Fermi II.

The panel members brought both prestige and experience to the work.[23] The Chairman was Mr. Robert O. Brown, president of the National Sanitation Foundation. Other members were Mr. Morton S. Hilbert, University of Michigan School of Public Health, Dr. Ralph G. Smith, director of the pollution and hygiene program, Wayne State University, Mr. Loring F. Oeming, retired executive secretary, Michigan Water Resources Board, Victor S. Sussman, president of Air Pollution Control Association, and director, Division of Air Pollution Control, Pennsylvania Department of Health.

THE CORPORATE CITIZEN

Introduction

An old American tradition was revived during the progressive movement, and was given articulate new expression by its leaders in both political parties. Theodore Roosevelt and Woodrow Wilson in their day, and indeed Hoover and F.D.R. in their day, insisted that the social and economic order was not immutable, and that the future was susceptible to modification, direction and even control, if men will but do it.

This became an accepted axiom of twentieth century America. The good citizen was the man who, at the cost of personal effort and private sacrifice, recognized the primacy of the public good, and strove to direct or accelerate the changes necessary to accomplish it. In a day when technology was triumphant, the possibility of directed change was hardly to be denied. The sometimes dramatic public confrontations and political conflicts were only over the question of how much direction was desirable, by whom and to what ends.

The application of the conception of good citizen to a great corporation would have seemed to the nineteenth century economic theorists an absurd contradiction of terms. Even the corporation, however, or at least its management, shared with these good citizens the concern for the future—and with results which were predictable. The Detroit Edison Company's plans for its own future came to involve so many people and so many public and private institutions that inescapably it acquired a public character. The great corporation, the most distinctive and creative economic institution of the twentieth century, by its size and by the obvious social consequences of its decisions, discovered that those decisions were not exclusively its own concern. Along with its privileges and opportunities, it also had responsibilities for the results of its actions.

Detroit Edison was and continued to be a business, with all the searching responsibilities implied, but beyond that were commitments which broadened with the years. Alex Dow had long argued that the interests of Detroit and The Detroit Edison Company were identical, but these statements had a passive tone. It was obvious that a healthy economy and a viable society had always been fundamental conditions to Company prosperity. His successors turned this tradition into a broader program for action.

The Detroit Edison Company was in no sense unique in this effort to define the disturbing new role, though the developing responsibility was not accepted everywhere so openly, or so frankly defended. The stockholders were reminded, in the annual report for 1967, of the three fold obligation which Alex Dow had so continuingly used to define his responsibilities. This was no longer enough. "Over the years we have emphasized our goals, to keep in balance the interest of the Customer, the Employe and the Shareholder. Now, management is faced with a fourth major consideration, namely, the public interest."

Social responsibilities and corporate considerations were thus extended to a concern with the general welfare. The Company, in so defining the obligations of the corporate citizen, had opened the road to remarkable corporate adventures.

No one planned it that way, and there was no self conscious social philosophy which gave explanation or defense or, indeed, which established boundaries for Company actions. In satisfying citizenship requirements, the Company found itself engaged in activities not common in utilities, and even stretching far from the lines of Edison's service area.

Occasionally alone, but more commonly with associated corporations, the men of Edison have joined or led the effort to summon business competence to public service. In these efforts, there have been disappointments, frustrations and even failures. On other occasions, there have been successes equal to the great effort.

Unhappily there is no statistical measurement for social responsibility, as there is for financial responsibility. The three chapters which follow are devoted to the circumstances of Company involvement in this significant area. No balance sheet is here attempted.

CHAPTER SIXTEEN

FERMI I AND THE FUTURE

The Company and Fermi / Proposals for Private Participation / Secrecy and the Amended Law / The Function of Fermi, and its Problems / The Struggle for Legal Existence / The Problems of Construction / Tests and Troubles / Success Within the Objectives / Administrative Leadership.

THE ENRICO FERMI ATOMIC POWER PLANT WAS FROM ITS INCEPTION freighted with the public interest. In 1951 when the first planning began, there was no public outcry on pollution, and no pressure group of conservationists. Smoke and the emission of particulates were old troubles, however, on which the Company had spent large sums in research and equipment. Nuclear fuels would end, not merely reduce, the nuisance.

Coal was in oversupply in 1951, and the price situation was favorable. Men who could look beyond that temporary situation knew that the national resource was being used wastefully, and that the supply was ultimately limited. If the nation which used half the world's developed energy thus destroyed its supply of fossil fuels, it was in peril—unless an alternative source of energy appeared in time. Nuclear fuel seemed to be promising.

The electric industry had of course an interest in both the availability and price of fuel. A great many utility and other companies shared in the cost of Fermi, in the hope of benefitting from the research conducted there. No one expected Fermi to be highly profitable as an operating power plant. But it was hoped that it would pay its own way, as men studied it and planned ahead.

In the research, and the development of the Enrico Fermi plant, the nation may have witnessed its largest single privately financed scientfic adventure, as certainly it was one of the largest efforts at industry-wide cooperation in the public interest. The leadership of the Company played a significant part in creating and directing this common effort, and thus for two decades the history of Fermi I runs parallel with the history of The Detroit Edison Company.

The knowledge of the power the atomic bomb implied intruded into almost any public or private consideration. The energy of the atom, if properly harnessed, promised to give mankind a potential of unlimited resources. The confusion of purpose alone made decision difficult, especially when digressive ends were so earnestly advanced by so many persons and organizations.

The electric power industry was among the interested parties, and in the year the bomb fell the Association of Edison Illuminating Companies named a committee to test the implications of these events. James W. Parker, president of Detroit Edison, was an active member of this committee. The members would have been happy to speak for the industry to a Congress considering legislation, but had only the slightest opportunity. There is no indication that their opinions and recommendations were more or less effective than any others.[1]

The Atomic Energy Act of 1946 contained a curious collection of contradictions. The preamble to the act announces its intention to strengthen free competition in private enterprise; the act provided for present complete governmental monopoly. The announced intention of promoting world peace was accompanied by rigid restrictions intended to keep atomic technology solely to America. Promoting the general welfare and an improved standard of living was a stated purpose, but the technical knowledge which only industry could have used was locked up quite out of its reach.

The Atomic Energy Commission, for which the act provided, could neither effectively use its information, nor could it allow others to have access to it. The transfer of control (January, 1947) from military to civilian hands did not modify the legislative non-discretionary provisions for secrecy.[2]

One of the contracting firms which had shared in the development of the nuclear bomb, Monsanto Chemical of St. Louis, undertook the development of an experimental nuclear power pile as a step toward commercial development. Monsanto invited James W. Parker to become a consultant in this project; formally, the contract for his services was between the two companies. Parker, as a member of the appraisal committee of the Association of Edison Illuminating Companies was highly interested in the problem, and in November, 1946, accepted the invitation from Monsanto.

It was April 9, 1947, however, before his security papers were cleared, and in July of 1947 the Monsanto Company announced it would not renew its contract as prime contractor for the Clinton laboratories. It was quite clear that Charles A. Thomas, president of Monsanto, shared the growing doubts about the possibilities of effective operations under the contractor system.[3]

In October, 1947, David E. Lilienthal, chairman of the new Atomic Energy Commission, made a speech to the Economic Club of Detroit, in which he restated the dilemma: the continuing need for secrecy, which was a legal requirement, made impossible the kind of open competitive initiative which characterized American private enterprise. On the same day, as if to emphasize a relation between the speech and the event, the Commission announced the appointment of its own Industrial Advisory Board, this time with official rather than volunteer status.

A considerable number of American industrial and business leaders were named to the Advisory board; the chairman was James W. Parker. Walker Cisler, who had recently returned to the Company, had been acting as consultant to the Atomic Energy Commission. With Parker's appointment to the Advisory Board, Cisler assumed the important post as executive secretary to the Industrial Advisory Board.

The Industrial Advisory Board worked from October, 1947, to the date of its report, December 22, 1948. The importance of the board is indicated by the attention its report received in the press, and by the thoughtful response which came from the chairman of the commission.[4] It was quite clear that, aside from discussions of staff organization, the proposals made were all related to the central problem.[5] But the Board still could not find a solution which would allow the Atomic Energy Commission to utilize the skills and resources of American industry, and, in turn, make its research available to American businessmen.

Independently but at about the same time (1950), Charles A. Thomas and Parker, representing their respective companies, approached the commission with proposals designed to break the deadlock. Each proposal involved the use of private funds which would make available to the Commission a supply of plutonium for weapon development without the capital cost of the plant required to produce it.

The development of Edison's program up to 1951 is not complicated. It was learned—at a social affair at the Parker residence—that the Dow Chemical Company might be interested in such a project. George Granger Brown, an engineer from the University in Ann Arbor who had been on leave with the Atomic Energy Commission, was also involved in the original discussion. Soon thereafter, Parker and Walker Cisler met with Dow officials in Midland for more formal consultations. Plans matured very rapidly then.[6] The planning group expanded to include Babcock & Wilcox, as well as some people from the staff of the Atomic Energy Commission itself. The Atomic Energy Commission announced on January 28, 1951, receipt of formal proposals from the Dow-Edison group, and also from Charles Thomas on behalf of Monsanto, and indicated its approval of the programs outlined.

What was proposed in each case was that the associated companies, at their own expense, undertake a study of the construction of an atomic reactor. The companies themselves and the men assigned to their responsibility would be cleared to handle secret information, and the commission would then make available to them its classified material and assign certain staff personnel as consultants. The details of such a program were complicated, and the completed contract between Dow, Edison and the Atomic Energy Commission was dated March 23, 1951.

The real purpose of this so-called Phase I was the establishment of the feasibility of the use of nuclear fuel in the economic production of power. Inside the Company, the Nuclear Power Development Department was established, with assistant chief engineer, Harvey Wagner, in charge, and Edward Nugent as his assistant. This was a supplemental assignment for both men. Working headquarters, with the essential security provisions, were established in Edison's offices.[7]

From the first, the Dow-Edison program had attracted enthusiastic interest from other companies and utilities, and industry in general. With the commission's approval of the plan, this interest widened, especially since at the moment the major cost of participation was supplying one or more men to the planning group.[8] By February, 1953, some ten companies were approved, and clearance was requested for another twelve; these chiefly represented utilities, but there were some manufacturers, including manufacturers of electrical equipment.[9]

The project moved smoothly from Phase I on feasibility, to Phase II, on plans and specifications. These were to be submitted in full detail to the commission before Phase III, construction, could begin. As Phase II was still in the area of theory and research, both the work and the expense were shared by companies across the nation.

The very progress in planning made clear the fact that the basic Atomic Energy Act of 1946 constituted an impossible barrier. If the companies in the private sector were to move in this field, they must be allowed to get possession of fissionable fuel, and develop the essential equipment for handling it.

Even before Parker's retirement as president, Cisler's driving leadership had shown significant results in this program. As president, he would speak with even more authority in his attack on the restrictive law.

Cisler was, of course, in no sense alone in these efforts. T. Keith Glennan, the president of Case Institute of Technology and a former AEC commissioner, suggested a conference on the matter which developed into the "Atomic Industrial Forum." The forum met in Detroit on April 16, 1953, with two hundred individuals and companies participating, and with Walker Cisler, chairman of the board of directors.

President Eisenhower further made a dramatic proposal for international development of the peaceful atom. When the Fund for Peaceful Atomic Development was organized for development and implementation of this proposal, Cisler was made president. These were worthy projects, and interest in them extended beyond their original purpose.

In 1953, the University of Michigan decided to memorialize its students who served in World War II with a program—called the Phoenix Project—to study the nuclear revolution in its many aspects. Since all the University's schools were involved, the Law School found itself with the responsibility for a study of the law as it related to atomic energy—a topic which the Detroit Edison Company found to be important enough to merit support. For a time, the Company's gifts to the Project were specifically directed to the Law School research. The Dean of the Law School, E. Blythe Stason, and his associates set themselves the task of examining proposals for the modification of the current Atomic Energy act. Research on this project took Dean Stason to many utilities across the nation, and to a number of conferences and conventions where the very nature of his inquiry emphasized inadequacies of current legislation.

In the spring of 1954, with an amendment to the 1946 law pending, Cisler arranged a meeting with a senatorial and congressional delegation. He took with him not only men from Edison, but also from Dow, Bendix and Ford. Cisler made no extravagant promises to this group—nor to any other to whom he talked in this period. No one could yet know with certainty either the technical or the economic implications of this new source of energy. He was insisting merely that the road should be cleared of the obstacles to development, and he emphasized particularly the importance of nuclear fuel as an essential supplement to coal or other fuels in the face of long term rising national energy demand.[10]

The debate increased in intensity through 1953 and 1954.[11] Unfortunately the efforts of the utilities to break the legal deadlock were interpreted as an effort to enable private exploitation of knowledge developed at enormous public expense. There were individuals in and out of congress who thought that the war related government monopoly should now be made the basis of a nationalized power system.[12] The new Republican administration was not in the mood to create such a system, although the debate was only in part partisan. It did leave the suspicion, however, that somehow the public interest was not being properly protected; this hostility and doubt showed itself again and again during Phase III of the program.

The revision of the law, which occurred in August, 1954, did not allow private ownership of fissionable material, but it did provide the means by which private firms could build plants and license fuels. With the legal foundation secure, the planning group increased its activity. In 1954, the budget for the work on nuclear energy for the so-called Dow-Edison project was $1,940,000, of which the Company furnished $320,000. The anticipated budget for 1955 was $3,800,000 of which the Company expected to contribute $750,000. By the opening of that year, there were thirty-three companies involved in the program.

One other important development occurred in the 1954–55 period. As the program developed in specifics, the divergence between the purposes and requirements of Dow Chemical and Detroit Edison became clear, and Dow announced its intention of retiring from the project to which it had contributed so much.[13] This action raised the whole question of overall organization.

On advice of counsel, and after conference with the associated companies, it was decided to create a nonprofit corporation. The letters of intent and agreement which had been the basis of the association to date were supplanted by a corporate form operating as a nonprofit organization under the law of New York (March 10, 1955). The overall organization was simple in form, but it proved confusing to some observers and to many newspaper writers. Atomic Power Development Associates, Inc. (APDA) was to be a continuing research agency of national scope.[14]

It was obvious however that the interest in the actual construction of a reactor would everywhere be local, and that a regional organization would be more acceptable for this purpose. The commission itself recognized the very great advantage in having several and diverse approaches to so new a field, and extended approval for planning to four groups. With slight exception, the members of these groups had participated in the research activity of the Dow-Edison project and the Atomic Power Development Associates.

Throughout 1955, plans were being matured for the development in the Detroit area. The creation of the Michigan corporation, Power Reactor Development Company, with some twenty-five members, distributed the corporate risk and the financial responsibility. The site chosen, south of Monroe, Michigan, had the advantage of being near the lines of several of the interested power companies. It was, however, within the Detroit Edison territory, and this fact tended to emphasize the Company's administrative relationship and responsibility. For good or ill, the new plant remained very closely associated in the public mind with The Detroit Edison Company.[15]

Early in 1955, plans for the plant were sufficiently firm to allow application for a construction permit. The Advisory Committee on Reactor Safeguards of the AEC made its examination, and very properly held that there was insufficient information on some items at the moment. The Commission ordered decisive tests before full power operation should be contemplated, and with this proviso, Power Reactor Development Company was granted permission to build at a site in Frenchtown Township referred to as Lagoona Beach.

August 8, 1956, was ground breaking day. Dignitaries from the city and state joined industrial leaders of the community and nation. President Eisenhower sent his congratulations in a personal letter to the Company, with the reminder that, just a year before, the first international conference on the peaceful uses of atomic energy had been held in Geneva. The chairman of the Atomic Energy Commission, Admiral Lewis L. Strauss, gave the main address, emphasizing the importance attributed to the plant and the innovative technology its plans proposed. The plant was named the Enrico Fermi Atomic Power Plant, honoring the late distinguished scientist (1901–1954) who first demonstrated the nuclear fission chain reaction at the University of Chicago, and who later had indicated his belief in the critical importance of the breeder reactor development.[16]

It is possible in a venture such as this one to overemphasize both the degree of innovation and ultimate significance of the operation. Most technological changes come through step by step advances rather than by dramatic breakthroughs. Journalistic reporting on the Fermi plant has been excessively influenced by the search for the dramatic and the sensational, with the result that much of it has been irrelevant.

A nuclear plant is for all practical purposes a conventional electric power plant. Water is heated to make high temperature steam under pressure, and that steam is directed through the turbines which power the generator. The heat in the traditional steam plants is drawn from one of the fossil fuels. The nuclear plant merely substitutes another heat source for coal, oil or gas.

Nuclear fuels are a source of energy which is neither the result of chemical reaction or of combustion, but evolves from the process known as fission. When the nucleus of an atom of uranium or plutonium is struck by and absorbs a neutron, the heavy atom usually will split (i.e. fission will occur) into two or more lighter atoms, and will emit radiation and neutrons. A neutron thus released can strike another atom of fissionable material and the operation thus becomes continuous. This is the so-called nuclear chain reaction.

A part of the energy released in fission is in the form of heat; a relatively small amount of fissionable uranium will produce a great deal of energy in this form. In 1964, the Federal Power Commission estimated that the fission of one-half of one percent of a seventy-five ton mass of nuclear material would, over a year, produce heat equal to the burning of one million tons of coal.

The intense heat from within the reactor is removed by a coolant, much in the manner in which the heat of an automobile engine is removed to the radiator by the flow of water in the cooling system. The difference, and a most important one it is, is that instead of using a radiator to dissipate the heat, the electric power plant conserves that heat to make steam and power. The coolant flows in a continuous stream from the reactor through the heat utilizing devices and back to the reactor.

The experiments and research of other groups, working with nuclear power and parallel in time with PRDC, developed several types of coolants, which served to identify their whole reactor complex. These included gas, heavy water, boiling water, and pressurized water. The common characteristic was that they used what is referred to as "thermal" reactors. In the thermal reactor, the speed of the neutrons released from fission is artificially reduced or "moderated."

Fermi planners from the first concentrated on the so-called breeder reactor, of which the most important distinction is the use of the "unmoderated" high speed neutron. With its associates in APDA the Company accepted, with this type of reactor, risks in complexity and innovation, along with what promised to be very great advantages.

The problems perhaps should be indicated first. The fuel used would be enriched uranium, in which the fissionable part is increased to about twenty-five percent of the total mass; the intensity of the reaction within the vessel is thus also increased. The coolant becomes a special problem, and at Fermi the decision was to use a molten metal, sodium, in the heat removal and transfer systems. One advantage is that sodium is not a moderator as water is. The use of sodium would allow the operation of the reactor at atmospheric pressure and this was desirable, but sodium introduced some problems of its own. With higher temperature, the operations were pushed close to the tolerance limits for some materials, including fuels themselves.

The advantages of the breeder reactor justified the most earnest efforts to solve these problems. The electric utility industry, conscious of the depletion rate of low cost coal reserves and anticipating a serious increase in the demand for power in the foreseeable future, had welcomed the suggestion of the use of nuclear fuels as an economic alternative. A second look, however, showed that the supply of uranium, or at least the proved reserve, was in no sense limitless, and might in fact be as restricted as was the reserve in coal. The facts as they developed showed that the conventional nuclear plant of the thermal type, such as was being built elsewhere, would be as wasteful of natural resources as was the coal furnace—though probably no more so.

It is at this point that the breeder reactor becomes important. The only significant element which is fissionable in its natural state is uranium. This metal is found in two forms, U 235, which is the fissionable element, and U 238, which is virtually inert. Unfortunately, the isotope which is active constitutes 0.7 percent of natural uranium. The prospects were for mining, refining and processing tons of ore in order to use a microscopic part of the total.

Fortunately the inert isotope, U 238, is what is called a fertile material, meaning that it can be made fissionable. So transformed, uranium becomes a man made element called plutonium, which is fissionable. In the operating reactor, some of the uranium 238 is transformed into plutonium. Some of that element will fission; that which does not can be recovered, but by a difficult process. This operation, in which, as a result of fission, fissionable plutonium is created and subsequently recovered for use as a fuel, is referred to as the plutonium cycle.

The tremendous advantage which both theory and experimentation attributed to the breeder reactor was that its high speed

The reactor vessel in transit from Chattanooga, Tennessee to the Enrico Atomic Power Plant.

The reactor vessel arriving at Fermi.

neutrons performed this transformation more successfully, and produced, relatively, a much greater quantity of plutonium. The rather incredulous result is that the breeder reactor produces more nuclear fuel than it consumes.

This, then, was the significance of the Enrico Fermi Atomic Power Plant. According to one expert who was a consultant in the planning stage, if the promise inherent in the breeder reactor could be realized, the effective use of uranium as a fuel would be multiplied one hundred times. Experience later reduced this calculation to forty times, which was still significant. The breeder reactor was important for a nation increasingly concerned with the problem of conservation, and the exhaustion of critical resources.

Since the Fermi plant was well within the Edison territory, any electricity produced would be distributed by that Company. The contractual relationship, thus, between Edison and PRDC, which owned the nuclear plant, was unusual and was modified at least twice. In the first change, Edison agreed to build and own the turbo-generator, powered by purchased steam from the nuclear unit. Ultimately, it was decided that Edison should also own the steam generator.

The entire complex was far from a laboratory experiment; the Fermi turbine generator, at 150,000 kw, was the duplicate of one then going into commercial operation at River Rouge. Although the Company was at the moment preparing to install some machines twice its size, the generator at Fermi was not small. The function of the Fermi plant was to translate the successful laboratory tests to the engineer's scale.

It was not expected that it would be competitive with a coal fired plant. This was a "first generation" breeder reactor and it was hoped that a second generation plant would be economically competitive—but only if Fermi proved itself. Both the limited size and the unusual distribution of financial responsibility and administrative control were designed to protect all participants, including the Company, from costly overcommitments.

There were three questions still to be answered, in spite of all the theoretical research and laboratory experimentation.

1. Could this plant be so constructed and operated as to satisfy an anxious public that proper safeguards adequately protected the adjacent area?

2. Could this plant meet the inherent demands of the economy, or show how they could be met? To do this it must show how power could be produced ultimately at a price comparable to coal.

3. Could this plant, and the attendant industry, perhaps ultimately with the assistance of government, solve the problem of the plutonium cycle and the development of its use as fuel? This only would bring to reality the possibilities seen in the breeder reactor.

The direct interest of the Company was expressed in these questions, of course, but much more they involved the future well-being and security of the nation.

The Atomic Energy Commission was properly concerned with the problems of safety. The permit to build issued in 1956 was not, of course, a permission to operate, which would come only on completion and testing of the equipment—which was estimated for December, 1960. The only condition laid down was that the Power Reactor Development Company prove within the year its financial competence to build.

Admiral Lewis L. Strauss referred, in his ground-breaking ceremony speech, to "some rather violently-voiced opposition in Washington. This opposition was not altogether unexpected by the Commission." The attack, he said, was "directed against the free enterprise development of nuclear power in this country, and in fact was against the provisions of the law itself."

The first "attack" came from a curiously distorted application of the Holding Company Act of 1935. The Detroit Edison Company had jealously guarded its independence from federal regulation, and a number of other utilities equally resolved were involved in PRDC; the proposals to extend full Securities and Exchange Commission regulation to the manufacturing companies and other business concerns because of their association with the Fermi venture seemed to be a straining of the law's intent. Even so, it seemed wise to move the whole administrative direction of both PRDC and APDA to a building on First Street in Detroit, in order to emphasize the "arms length" relationship.

Because the Fermi proposal was the first of its kind, it received full opposition from the advocates of public power. The twin corporations advancing research and development for the plant were attacked on the floor of the Senate as tax dodges, and because payments by the member companies were listed as expense for tax purposes, the operation was referred to as a "subsidy of the taxpayers money."

The most overt attack came in connection with the bill introduced by Senator Gore and referred to the Joint Committee on Atomic Energy. The Senator proposed the establishment of six federal atomic-power electric plants on a regional basis, and defended the suggestion with the assertion that nuclear development in the United States was lagging. Cisler, in his testimony, called the charge "the product of a fertile and purposive imagination." He pointed to the fact that ground would be broken at Monroe one year after permission was given to plan a plant, and insisted that all available man power and scientific resource was being utilized on the program at the moment. He argued that a federal crash program would only scatter or at best transfer these efforts.[17]

The Gore bill was defeated, but Congress adjourned with the two bills in which the industry was interested still in committee. They included a bill for a federal disaster insurance program, and a measure which would specifically rule that PRDC was not a holding company.

The worry about the tax status of contributions to APDA and PRDC, and about the possibility of Securities and Exchange Commission action relative to the holding company, came unfortunately at the moment when financial matters were being decided. Contributing companies might well worry over collateral effects. Cisler had to urge strongly that the member companies refrain from making grants conditional on one or another of the desired laws or rulings.

Cisler presented the financial situation as he then saw it to the board of Detroit Edison, on October 15, 1956. The plant would cost an estimated $43.5 million, including $7 million research to be done and paid for by APDA. Arrangements had been made to borrow a sum of $15 million from banks, with the loan guaranteed by the several companies. Direct contributions from the member companies would amount to $23.79 million. The Detroit Edison board agreed to raise its earlier commitments to a total, past and prospective, of $8.75 million and an underwriting guarantee of $9 million of the bank loan. It was expected that the costs of this borrowed capital would be met from the sale of steam and plutonium. The Michigan Public Service Commission gave its approval to this financial arrangement. Detroit Edison participation was spread over a five year period, and most of the other companies likewise limited the extent of the commitment.[18]

The problem of safety which became so critical for this particular nuclear plant was introduced into the whole discussion almost by accident. At the hearing before the Atomic Energy Commission on the application for permission to build, the Advisory Committee on Reactor Safeguards had presented a negative report, insisting that there were as yet unanswered questions. In granting permission to build, the Commission, aware that extensive tests were planned following construction and preceding any consideration of an application to operate, in a sense disregarded the advice of its own committee.

Commissioner Murray referred to this episode before the closed session of the Congressional Joint Committee on Atomic Energy, and was quoted on the Senate floor by Clinton P. Anderson as saying there was insufficient information available at this time to tell whether or not this reactor could be run at this location without public hazard. In so quoting Murray, without reference to the scheduled testing program, Anderson started quite a furor in the newspapers.

At the height of this excitement, three labor unions moved to intervene, claiming to represent a public interest. They charged that the requirements about financing had not in reality been met, and they demanded that the commission refuse permission to build until all questions of safety had been completely resolved. Governor Williams of Michigan had designated a state commission on Atomic Energy under chairmanship of John H. McCarthy, of the Public Service Commission, to make an investigation on behalf of the state, and was reported satisfied as to the safety of the plan. Local governments, including the city of Monroe, urged construction.

The three unions were the United Automobile, Aircraft and Agricultural Implements Workers of America, the International Union of Electrical, Radio and Machine Workers, and the United Paper Workers of America. They requested a hearing and asked that construction be stopped in the meantime. The hearing was granted but the permit to build was allowed to stand. The hearing was set for November, but then postponed to December 10, 1956.

That December and January offered illustration of a new national business climate; Detroit Edison was in Washington on three different hearings, all concurrent.[19] It was asking the Federal Power Commission for permission to increase its power exports to Canada. A shortage there was the indirect result of the rock slide which destroyed a power plant at Niagara Falls. Permission was granted. The Company was asking the Securities and Exchange Commission that PRDC be held not to be a holding company. The SEC so ruled.

The intervening unions' hearing before the Atomic Energy Commission finally opened on January 8, 1957, on the request to reconsider the construction permit. The associated companies responded to the charge of inadequate finance, by increasing the duration of the annual pledge from five to six years. On the problem of construction and safety, PRDC supplemented the testimony of its own staff with that of APDA, and also brought in Dr. Hans A. Bethe, nuclear physicist and at a later date Nobel Laureate from Cornell, Dr. Norman Hilberry, Argonne National Laboratories, and W. Kenneth Davis, Atomic Energy Commission Director of Reactor Development.

The final ruling from the commission came in June, 1959, and served to confirm the decision first made in August, 1956. This proved not to be the end of the harassment. The unions, through counsel, appealed the decision to the courts, and thus invited further inevitable delays. In June, 1960, the Court of Appeals ruled that the Atomic Energy Commission did not have the right to allow construction until all questions relative to safety had been completely cleared.

The three man court divided two to one, with the minority judge, future Chief Justice Warren E. Burger expressing vigorous dissent. He said that "from an erroneous premise drawn out of thin air the majority had proceeded to draw an unwarranted conclusion."[20] The decision would seem to have the effect of making the permission to build the equivalent to the permission to operate.

The Detroit News, commenting on June 14 on the issue and the decision, quite bluntly charged that some leaders, including Walter Reuther who was named specifically, were attempting to use this as a device for blocking the entrance of private business into the nuclear field. Safety, in the editor's opinion, was only an issue of convenience.[21]

PRDC took its proper appeal to the full nine man Court of Appeals, and in this action found itself joined by the Atomic Energy Commission itself. The three man court, in the language of the appeal, had substituted its opinion on safety for that of the properly constituted authority on the subject. The commission was concerned, for the effect was to frustrate its power to regulate. When the nine man Court of Appeals did not reverse the decision, PRDC filed a petition, August 12, 1960, asking for a hearing by the United States Supreme Court.

By this time, the case had become constitutionally significant, and both the Solicitor General of the United States and the Atomic Energy Commission were insisting formally that public policy was involved.[22] On November 14, 1960, the Supreme Court agreed to review the Court of Appeals decision, and it was agreed that the government case and that of PRDC would be consolidated in the pleading. The decision (7–2), dated June 12, 1961, upheld the right of the Atomic Energy Commission to make decisions, and thus the decisions which it had already made.[23]

Five years before, when PRDC and APDA had celebrated the ground breaking, Cisler said, with what seemed then to be extreme caution, that Fermi would be producing electricity by 1961. Instead, they had in 1961 a legal confirmation of the permission to build. But the ten years since the completion of the feasibility study, and the five years since the receipt of the original permission to build had not been wasted. The Supreme Court's decision found the Fermi plant close to physical completion.

Despite legal distractions, building had been almost continuous since 1956. In the nature of things, this construction was slow, for the essential equipment had to be invented, designed and built, and the installation of unit parts must be followed by tests for compatibility and safety. This was the first plant of its kind and size in the world.

The reactor vessel itself was fabricated into a complex weighing ninety-one tons. It was floated on a barge from the factory at Chattanooga down the Tennessee river and up to Cincinnati. Then a route had to be devised for the especially built railroad car to transport it to Michigan. It was shipped by way of Belfast, Indiana and Grand Rapids, Michigan, and even on this special route, bridges and underpass supports were occasionally cleared with only an inch to spare.

There were two new components which demanded special security testing. From the beginning, the decision to use liquid sodium as the heat exchange fluid had been recognized as a hazard which must be accepted. Sodium, in this form, reacts violently when exposed to water. Literally years were spent studying this reaction, and developing means and materials to control the results should an accident occur. Equipment to pump and control the flow of the metal had been tested on small scale, but no one could be sure of the results at operating size.

The other novel item was of course the reactor itself, in which radioactivity and intense heat combined to pose new problems. The mechanism for handling the fuel, the insulation acceptable in extreme conditions, all the complex safety devices up to and including the total "scram" operation were the product of extensive and often inventive efforts.

Construction was in the hands of the Power Reactor Development Company, which had been formed as a separate corporation, totally distinct from Atomic Power Development Associates, the research organization. Once Fermi started, however, the research activities tended to move to the building site. APDA concentrated its attention on the problems of sodium heat transfer and fast breeder reactors; at the moment the nation's activities in these matters were centered at Fermi.

It was necessary to establish a kind of arbitrary division on function and cost between APDA and PRDC. PRDC was assisted in the elaborate development and testing program by the AEC under terms of a contract dated March 26, 1957. Some of the testing required equipment which only the commission had available. The contract stipulated that the commission provide assistance in research up to $4,450,000. The service included cooperation in training operators. Under the law, the commission must retain ownership of the fissionable materials, but under the terms of the assistance contract, these materials were leased for a five year period without charge.

One of the requirements imposed was that of reporting progress to Congress, in an annual statement to the Joint Committee on Atomic Energy.[24] In March, 1961, Charles Landrigan presented the statement on behalf of the Company, with special reference to the nonnuclear parts of the structure. He reported that the turbogenerator, manufactured by Allis-Chalmers of Milwaukee, was in place, and was rated at 156,000 kw, and that the whole structure had been substantially completed by December, 1960. Landrigan was still estimating the costs of the nonatomic installation at $17 million, totally at Company expense.

The termination of the construction period was at hand, and changes in administration recognized that fact. Robert W. Hartwell had for a time headed the Nuclear Power Development Department within the Company, and had served briefly on loan to APDA. Since 1957, he had been with PRDC, briefly as assistant general manager and then as general manager, during which time he had been on leave from Detroit Edison. As of December 31, 1963, he returned to the Company; John Anderson had been assistant general manager for almost the same period, and he too returned to regular assignment.

For a number of years, Arthur Griswold, vice president of the Company, had been in charge of relations with Fermi Power. As of January 1, 1964, Griswold succeeded Hartwell as general manager, and was also elected vice president of PRDC. Walter J. McCarthy, Jr. had been with the APDA research group from the beginning, and had later moved to PRDC. In September, 1963, he became an employe of Detroit Edison, on leave to PRDC. He had been and continued to be assistant general manager. Until 1968, the direct operations of the Fermi plant and of PRDC were in the hands of Griswold and McCarthy.

Any power plant is an exceedingly complex affair, and normally up to a year can be needed to complete the corrections which actual operation shows to be necessary. Fermi had perhaps an undue share of these so-called "shakedown" problems, most of which were revealed in the early nonnuclear testing period. Unfortunately, careless newspaper reporting merely identified them as accidents at the nuclear plant.

The nonnuclear testing had been scheduled to occupy a period up to June, 1961. Three major failures did develop—all mechanical rather than atomic—which had to be remedied sequentially, thus extending this part of the testing program far beyond the calculated time.

As a part of the testing, temperatures were raised to 1000 degrees F (the margin above the expected operating temperatures of 900 degrees F was not excessive). During the test, the graphite shield, which had been especially devised to meet the demands of combined heat and radiation, broke down. The removal and replacement of the entire installation had to be undertaken, and the problem again thrown to laboratories for research. The time lost before new materials were installed was from December through the following summer, 1962.

An even more critical source of delay was trouble with the fuel handling equipment. The dummy fuel elements had been installed in the reactor to simulate actual conditions when heated sodium was being circulated. In the nonnuclear test period, as a result of an operational mistake, the dummy fuel assemblies and the fuel support systems were damaged.

This meant the redesigning and rebuilding of the entire system of fuel supports. In the meantime, the mechanism had to be drained of the metallic sodium—more than 345,000 pounds of highly dangerous material—in order to reach and dismantle the defective and damaged parts. Safety required the exclusion of air, so the task was performed in an atmosphere of argon gas. The men working inside the vessel wore uniforms prophetic of the future walk on the moon.

The third problem was on the Detroit Edison side of the Fermi complex. The sodium flow provided the heat which Edison would purchase. The boiler which used that heat to create steam was a constant problem with difficulties from design and fabrication. Since the water tubes were in a bath of the sodium, small leaks or other imperfections, which would have been tolerated in the standard boiler operation, produced dangerous consequences here. A number of tubes in one bundle ruptured setting off so violent a reaction between water and the sodium as to be momentarily critical. The safety devices designed for this emergency worked well, which was some comfort. It was eventually determined that the failure was the result of excessive tube vibration during operation and this was corrected but the boilers continued to be vulnerable to troubles.

These failures added to the already considerable delays in the ultimate nuclear testing. In 1958, Hartwell had expected to move from low power operation to full power production between September, 1960, and September, 1961. At the end of 1961, this estimate was sharply modified in the annual report, as these mechanical corrections could not be completed before the end of 1962.

On October 11, 1962, the Advisory Committee on Reactor Safeguards reported to the Atomic Energy Commission that in its opinion Fermi could be operated for the requested test period at one thermal megawatt,[25] and public hearings were ordered for January 3. Again there were those, claiming to represent the public interest who tried to intervene, but the commission approved (May 10) the recommendation of its committee. The Enrico Fermi plant went "critical" at 12:23 P.M., August 23, 1963. Testing was to proceed at least six months before application was made for permission to move to full power.

On November 17, 1964, the committee again recommended to the Atomic Energy Commission that a request for permission to operate at 200 thermal megawatts be approved. Final application, together with the results of up-to-the-minute tests, went formally to the Atomic Energy Commission April 23, 1965. The amended operating license, dated December 17, 1965, gave the requested permission, and almost at once power intensification was increased in planned stages. It was intended to take about six months to bring the unit up to the anticipated full power.[26]

These were exciting days, for Fermi's potential was now increasingly understood.[27] The first nuclear operation was followed shortly by an important conference on the breeder reactor (December 2–3, 1963) in which two hundred scientists, utility specialists and public figures participated. There was a constant stream of visitors to the plant, including many from abroad.[28] The British had built a small breeder reactor in Dounreay, Scotland, and proposed now to build a larger one; Fermi was, by their admission, their intermediate step. A European international organization (Euratom) proposed to use the Fermi achievements in their second generation project. Even an important group of Russians, including both scientific and political figures, made the visit to Fermi an important part of their tour of the United States.

In August, Walker Cisler sent telegrams to the directors of the two companies, and to the directors of Edison, reporting that on Friday, August 5, 1966, Fermi had operated at one half its rated core power level, and continued to do so over a fifty-five hour period, August 5–7. In the period of the test, Fermi had produced one million kwhr of electric power.[29]

Fermi's success was not yet complete. On October 5, 1966, an incident occurred which closed down the operation of the plant completely. Two small triangular pieces of sheet metal, a belated precaution against one kind of malfunction, broke from their fastenings under the constant sodium stream vibration. The strong sweep of the sodium carried one of them along to lodge in such a position that it obstructed the flow of the coolant to several of the fuel elements. The temperature in the affected part rose above the established safe level; some of the fuel pins melted and distorted their metal supports.

There was no panic and no injury, but opponents of the plant used this incident widely to reinforce their continuing effort to block its operation. This was a mechanical failure, the result of a construction decision not totally understandable. Fermi remained closed down for more than three years, while efforts were made to right the difficulty.[30]

It would not have been as unhappy a situation for the Company if earlier predictions had been less optimistic. Previous forecasts and timetables had not anticipated fully the delays in development of technology or equipment. Other utility groups, some also behind schedule, completed their water cooled reactors and announced operation of their nuclear plants while Fermi still struggled. The very real achievements, when so delayed, somehow lost their impact, even though Fermi had in action substantially confirmed its theoretical conception.

The basic purpose of Fermi, the one which justified the investment of funds by twenty-two other companies, was research and innovation. The articles of incorporation of PRDC made the purposes quite clear. "To be a business league of members of industries—interested in the conservation of natural resources and the economical production of electrical energy, and to study, develop, design, fabricate, and construct and operate one or more experimental nuclear power reactors and any and all component parts—." In the decade which followed that first statement, the research objective had been constantly emphasized.

By the summer of 1964, however, when the application was made for permission to increase the level of operations, the situation had changed in two important regards from that which held when Fermi was planned. The first had to do with the law under which PRDC operated, which required that plutonium be sold only to the government.[31] Under the new law (August 26, 1964), the most important amendment to the law on atomic energy, the government announced its abandonment of the requirement; this meant the end also of the supported price. Even with support, the price of this material had fallen to about one fourth the original price. The effect on the economy of an operating Fermi would be significant.

The second change was also a product of the law, the provisions of which were scheduled to become effective over a period of a few years. The system of government monopoly and leasing of fuels was abandoned; even though the significant enrichment program remained in government control on a toll basis.

In the period when all plutonium was controlled by the government, the government was concerned chiefly with weapons. Government did not, and private industry could not, perfect an industrial (as opposed to a laboratory) method of converting the material to its potential use as fuel. Until this was done, the breeder reactor had to remain a laboratory experiment rather than an industrial breakthrough.[32]

These things being true, the Enrico Fermi plant, and especially the reactor, had other things to do more important than the production of electricity. In January, 1964, before the nuclear tests were completed, the Company ordered an oil fired boiler to be used in conjunction with the Company owned turbine generator already installed there. It would provide steam during those periods when the Fermi reactor was closed down or was operating for research purposes. Commenting on the Fermi situation in the annual report for 1964, Cisler again made the point that the new law of that year raised the whole question of the best future use for Fermi.

Support continued to come to the research program from the member companies, but it was hoped that the research program could become self-supporting. PRDC accepted contracts for tests and research from the Atomic Energy Commission, from Euratom, and of course, from the member companies. By 1966, Fermi was within sight of this goal.[33]

The question being asked at Fermi now was no longer whether

or not the breeder reactor would work, but whether or not the economic advantages which theory anticipated could in fact be realized. This was now a problem of fuels, enrichment programs, heat transfer components, handling equipment, design, and utilization, and for the vital research in these fields, Fermi was unique in the world.

There are still unanswered technological questions in the operational field of the breeder reactor. But inside Detroit Edison, there is still the conviction that atomic energy is a national essential, and that the breeder reactor is the method by which that power will ultimately be attained. The year 1980 passes currently as the date when commercial application can be expected.

Almost lost to view in all the concern with nuclear potentials and the technology of fission control is the administrative and financial means by which the whole was accomplished. In 1955 the largely informal arrangement between Dow and Edison and the associated companies gave place to the two distinct corporations. For present purposes, the separation is of no importance. The cooperation between the two was continuous, there was some exchange of personnel, and Walker Cisler as president of both corporations gave them a common direction.

In the decade which followed, these two corporations organized, financed and directed a research and development program which, devoted as it was to a single project, is probably the largest such private activity in world history. For a basis of comparison, one finds himself turning to such governmental projects as the one which created the original nuclear bomb.

Before 1955, in the days of the feasibility study, representatives from the member companies served the project either full or part time; even in these early days, as many as 188 individuals selected from many companies worked at this job. After the formation of the two corporations, APDA and PRDC continued to have access to the personnel of the highly diversified member companies.

To this significant technical resource must be added the staff recruited and developed over the ten years of activity, and employed directly. The list of persons who were employed as consultants is a roster of the world's leaders in both nuclear theory and accomplishment. The Atomic Energy Commission's contribution included staff services and expert consultation.

The scientific achievement was no more remarkable than the administrative one on which it depended. Cisler's enthusiasm and vision pulled together companies of diverse backgrounds and interests, held them in the face of adverse publicity and frustrating delay to the support of the program and then to its extension.

Atomic Power Development Associates was a research organization, nation-wide in its intended reach, and in the strict or legal sense related to Fermi only in an advisory capacity. The forty-three companies, which at some time contributed, understood from the beginning that money which went to APDA was a contribution to the development of the science. This was not so clearly stated in connection with Power Reactor Development Company, which was to build the Fermi plant and own it; this company ultimately had twenty-four members.[34]

Beyond question, it had been the hope that the company building Fermi might be able to produce energy at a price which would allow some return on investment. The nature of the organization, however, seemed to make explicit the fact that such return would not be guaranteed; the inclusion of twenty-four corporations in the membership was an effort to spread the economic risks. From the beginning, the Enrico Fermi plant was owned by a corporation specifically listed as "nonprofit" under the law, and the payments to it by the corporate members were referred to in its annual reports as contributions.

The total cost of the Fermi project has been variously reported. The estimate of $120 million, which has been the journalistic approximation, is only roughly accurate. The question both of what it cost and who paid for it ought to be rather factually reported, and this involves recognition of changes as they developed.[35]

Through 1961, when the plant was completed and the tests in process, the financial commitments for the three related enterprises were quite simple. To that date, research costs of APDA were $27.8 million. The Detroit Edison Company owned and would operate for its own benefit the turbine and the steam generator, on which it had spent $17.2 million. Neither of these was a part of the Fermi nuclear development. The actual costs to date of that reactor plant were $49.7 million.

That cost had been met by commitments by the twenty-four member companies to the total of $35.2 million; of this amount, $27.5 million had been paid in cash, and the rest was in valid pledges. The balance of the expenditure was covered by a bank loan of $15 million. It was clear that, if plans had materialized this loan would have been financed from income; the group of banks, however, which made the loan insisted on endorsement of specific amounts of the loan by companies as a direct obligation. Thirteen companies had underwritten a share of this loan, including interest and principle. The notes became due from 1964 to 1970. Of the $35 million pledged and/or paid to December, 1961, Detroit Edison had paid or pledged $14 million, which is 40 percent of the total.

Between December, 1961 and December, 1968, the financial picture changed. The elaborate series of tests, the repairs following the 1966 breakdown and the closing of operations, together with the rise in all prices, resulted in a sharp increase in costs, uncompensated by any operating income. The participation of APDA was increased, chiefly as a result of the fuel meltdown, from $27.8 million to $32.5 million; Detroit Edison's addition of the oil fired boiler was its chief increased investment in the Edison owned components from $17.2 to $22.9 million. PRDC's costs for construction, and including operating losses, rose from $49.7 to $66.2. These three figures do indeed exceed the reported $120 million, but the PRDC item is the significant one.

The financing of this cost increase was difficult. Cisler, as president of PRDC, was effective with the board of Edison and the member companies in his appeals for additional commitments. The end of the year 1968 found old and new pledges paid completely, with sixteen of the twenty-four member companies sharing the increase beyond the original commitment. The amount so pledged was increased from $35 million to $50 million. In the absence of supporting revenue, the bank notes had long since become an obligation of the endorsing member. The thirteen companies which had underwritten the bank notes had paid $14 million in principle and interest, with only a small balance outstanding. This contribution of $65 million by so many companies is a testimony to their faith in the prospects of the breeder reactor, and their continued confidence in and support of the leadership in this prolonged adventure.

Yet there was a note of finality in the PRDC report for 1968. Pledges had been paid, but it was obviously the end of this kind of financing. In the words of the annual report (May, 1969) "Regular operations beyond the middle of 1970 are dependent upon many factors, including the obtaining of additional funds, and therefore, are not determinable at this time."[36]

Some persons will insist that projects of this dimension are properly left to the hands of government, and that the financial expenditure—avowedly for the national benefit—should have been shared by a tax upon the larger community. It is certainly true that at least some participants, if they had foreseen the time and cost involved, would have questioned their first commitment.

Detroit Edison was asked to share a disproportionately large part of the unanticipated increase in costs after 1961. Its share in Fermi's nuclear experiment went from $14 million in 1961 to $22.7 million in 1968, to which must be added the $7.9 million paid on the guaranteed bank notes. The total paid by Edison was thus $30.6 million, which changed its share of the total project cost from 40 percent to 47.5 percent.[37]

Costs, however, must always be relative. The contributions to Fermi, charged to expenses over the years, was well under one percent of the Company's expenses. If this nuclear development is recognized as a long range contribution to conservation and the protection of environment, then the total Company expenditure finds a relevant comparison. Cisler's estimate on the other expected costs of pollution control,[38] recent and prospective, was twice the amount the Company spent on Fermi's nuclear plant.

Despite the project's detractors, neither the safety of the city nor the solvency of the Company were in the least endangered by Fermi.

CHAPTER SEVENTEEN

THE CITY AND ITS NEEDS

The Corporation and the Public Interest / Gifts and Grants are Easy / Organized Civic Advancement / Urban Problems and the MDCDA / The Doxiadis Study / The Summer of 1967 / The Riot and The Reaction / At Our Own Back Door.

THERE IS NO MAGIC WHICH CAN DIVIDE THE COMPANY FROM THE AREA IT serves, or allow either a separate existence. The dependence of the Company on the continued economic health, the social responsibility and the peace and order of the greater community is a reality. The obligations of a concerned citizenship thus tended to merge perceptibly with Company interest.

It was obvious that the Company could not successfully plan its own future in an urban complex which was itself planless. The service area of the Company extended to cover southeast Michigan, but roughly half of its revenue came from Wayne County; the headquarters of the Company was located in what was being called the "inner city." Such facts emphasized the self evident involvement, but there were the decisions to be made on degree and means of this participation.

In the earlier years, the decade of the fifties, the community spirit of Detroit found a most interesting form of organized expression. Probably the type of interindustrial civic committee is not unique, but there it was institutionalized and became highly successful. In this period, Edison was one of a large number of participating corporations.

In the troubled days of the next decade, men from Edison led in the effort to reestablish that social concern, and thus to bring to bear the talent and resources of the business community for the cooperative approach to an urban crisis. The civic committee device was again activated in the effort to achieve cooperative action, this time with the membership extended to include other segments of the community. This chapter will discuss the development of the method, the efforts of the men of Edison to revive and redirect the community effort and their participation in it, and finally the efforts of the corporation itself as a resident of the inner city to deal with its own neighborhood.

Edison had long recognized the importance of being a part of the several communities which it served. As early as 1953, the annual report coupled, as obligations upon employes, service to the Company and "improving the social and economic environment in which it operates." Edison men were encouraged to offer assistance, especially their technical competence, to governmental agencies, and they served on boards, commissions, and authorities across southeast Michigan. In local organizations of public or civic character, membership was often at Company expense.[1]

Money contributions to worthy causes became an accepted part of corporation practice. World War I gave justification for financial contributions to war-related and relief agencies, and established a legal precedent more significant than the mere money involved. The management of the corporation was permitted to use "stockholders' money" for purposes other than dividends. In so doing, management was granted a limited discretion in other expenditures as well.[2]

In some cases, there was an obvious Company interest expressed in the gift. The contributions to educational institutions on the professional level were generous, as were the grants to support research. It would be hard on some occasions to separate such grants from the efforts at recruitment of persons with specialized skills. But this was not always true.

The pressure for gifts to educational institutions led to the creation of a special board committee on Aid to Education and to the proposal that such education grants be limited to one half of one percent of income before taxes.[3] This decision provided for increasing amounts as income rose. Eighteen Michigan colleges without tax support were especially recognized. By 1965, the amount budgeted to education exceeded one quarter of a million dollars.[4]

In the main, both in the Detroit area and in statewide and national appeals, the greatest care was exercised in making other grants. On the local scene, when an organized community campaign was launched, Edison could expect an appeal for financial assistance. When such contributions were recognized as a legitimate part of Company expense, then the only decision was how much and to whom.

When Detroit celebrated its 250th birthday in 1951, it was decided that a committee should be formed to accept contributions and select a suitable gift to the city. The Capital Gifts Committee was incorporated for this purpose,[5] and two structures resulted. The first was an attractive building devoted to the Arts on the campus of Wayne State University. A much more elaborate structure, Cobo Hall, was begun as a part of an impressive water front development recently begun, and was completed in 1960.

So unusual and so successful a civic undertaking suggested making permanent the means of its accomplishment, so the Gifts Committee became the Detroit Conference on Civic Development. Walker Cisler was not on the original board of incorporators, but in 1954 he was elected to the board and became its cochairman at the meeting at which the name change was approved.[6] The annual operating budget, as much as seventy-five thousand dollars, was contributed by various industries.

The Conference assumed the continuation to completion of gifts proposed to a total of $9 million. It also accepted additional responsibilities, of which the chief was administered by the Office of International Relations, under the direction of Ruth E. Stevens, vice president. Mrs. Stevens had general responsibilities, but was especially in charge of civic functions for visiting foreign dignitaries. At the end of ten years, Mrs. Stevens reported functions honoring thirty-eight heads of state or other high dignitaries on official visits from twenty-five countries.

A new committee under the Detroit Conference was the Detroit Tomorrow Committee. This was in response to Mayor Cobo's request for a representative group from the business community with whom departments of the city and city hall itself might consult. In only a short time, the responsibilities of this committee became greatly broadened.

In the erratic economic climate of the fifties, in which swings of production and employment were matched by equally unpredictable population movements to and from the city, its economic leaders were concerned about the future. Especially after the economic slowdown of 1957–58, a rash of committees, organizations, and associations developed to save Detroit and Michigan. One employe, asked to represent the Edison business organization on such a committee, reported a confusing number of volunteer efforts, and Cisler counted six organizations given official standing by one or another of the levels of state government.[7]

It seemed essential that the various volunteer, sponsored and official organizations be in some matter coordinated. To that end a group called "The Mayor's Committee on Economic Growth" was named, the membership of which was widely representative.[8] Walker Cisler accepted the appointment as chairman. This committee superseded the older and smaller "Tomorrow Committee," though it continued to function as a part of the Detroit Conference.[9]

The results of the activity of the committee members, coinciding with the upturn of the economy, brought to visible result the civic planning of a decade. As early as 1948, Detroit had adopted a master city plan; for a long time it was the only city among the nations' five largest with such a conceptual framework. Ed Jeffries, as wartime mayor, had accumulated a small financial backlog for this purpose. His successors followed his lead, and the city began a downtown and water front modernization which was indeed impressive.

The organization of community effort, and the mobilization of corporate resources in men and money brought outstanding results in a second effort in the Detroit area. The Metropolitan Detroit Building Fund substituted for individual and somewhat duplicative campaigns for money a program of pledges payable over a series of years, and allowed the allocation of these funds to approved organizations in an orderly and assured program. More than $14 million was distributed to build or extend nine hospitals, and capital gifts were made to twenty-one other philanthropic or service organizations. The Metropolitan Detroit Building Fund succeeded and extended the community activity of the fifties. Edison shared costs in this program with other corporations of the city, and Edison men helped plan and organize the campaign.[10]

The wartime and postwar industrial demands for labor had stimulated a migration to the city from rural America. Both blacks and whites were involved in the massive movement from the south, but in Detroit, as in many other northern cities, the proportion of the population which was referred to as nonwhite increased sharply and continuingly. Growing in a kind of formless sprawl, the city itself became a problem.

As the newest immigrant, the Negro fell heir to the cheapest and least desirable housing. This defacto segregation was as much a product of economy as it was of race. The skills of the agricultural worker were not directly transferrable to the industrial world, and almost certainly the move to the city would be followed by severe economic difficulties. For many, this period was short.

As rapidly as acquired habits and skills brought permanent jobs and adequate income, members of the black community began a movement outward from the slum area, but only to have their places taken by a new group of migrants. The slum, to all appearance, had become a permanent part of the urban scene; what was even more alarming, it seemed to be expanding. Both the federal government and the city were moved to do something about so conspicuous a disorder.

Subsequent frustration and criticism should not hide the fact that massive efforts had been made to deal with this urban problem. In Detroit, the federal government and the city had combined to commit one quarter of a billion dollars in subsidy, and private industry had invested almost one hundred fifty million dollars more in construction in the cleared area. The citizen might well feel that these were impressive sums, and that more should have been done with so much.

Slum clearance, at public expense, was expected to provide land at a price which would subsidize approved private construction. The effort to involve the private sector of the economy in a more massive effort was not broadly successful. The theory was good, but restrictions, confusions, inexperience and mismanagement combined to leave only bulldozed rubble on block after block where houses had stood. They had been modest houses, perhaps even substandard, but the important fact was that they had not been replaced.[11]

In considerable part, the problem could be reduced to one of price. Even with the land at a low cost subsidy figure, the cost of construction of new buildings at the higher required standards made either the rent or purchase of these new residential units beyond the financial reach of those persons who had been displaced.

The slum clearance and redevelopment program as it had evolved over two decades obviously needed critical reconsideration. At almost the same time in the city and in the nation's capitol there was evidence of a desire for a new start.

President Lyndon B. Johnson, in his 1966 message to Congress made a proposal which, with several intervening changes, ultimately became the Model Cities Program. Even before that date, conversations involving Jerome P. Cavanagh, Mayor of Detroit, Walter Reuther, president of the U.A.W., and Walker Cisler had indicated a desire to act.

While the Johnson proposal was still in debate in Congress, these three men moved to establish a basis for action in Michigan. They became the incorporators of Metropolitan Detroit Citizens Development Authority (MDCDA), a nonprofit corporation. The incorporation dated from November, 1966, and at an early meeting the three incorporators designated a group of directors on whom long term responsibility would rest.

The purposes of the incorporators were obvious. This was to be a massive attack on the problems which, because of confusion, cross purpose and inadequacy, had frustrated early efforts by groups and by individuals. The organized community, extended this time to include a six county area as the metropolitan region, was to be involved, and the driving force of business leadership together with the resources of the business corporations were once again to be mobilized in the public interest.

The significant new element is the fact that MDCDA did not place sole dependence on housing for the solution of the problems of the urban blacks. The effort to involve other elements of the society than business, to create community cooperation on the local level, and to create a sense of individual and collective responsibility among the people concerned all suggested a new approach, the details of which were yet to be developed and tested. Unfortunately, time ran out for Detroit in the summer of 1967.

In 1948, when Cisler was in Greece as a working advisor with the Marshall Plan economic program he met and worked with the Minister-Coordinator of the Greek Recovery Program, Dr. Constantinos Doxiadis. The two men became friends, and in the years which followed, their mutual interests drew them together on a number of occasions.

Following the war and the reconstruction program, Doxiadis achieved wide recognition as a planner and urban developer. He founded a private firm of consultants, Doxiadis Associates, with headquarters in Athens, but with subcenters in several locations, including Washington. His services were used by the United Nations, by the International Bank, by a dozen countries in Asia, Africa, and south Europe and by private concerns around the world.*

Walker Cisler had long been concerned with the problems of planning inside the electrical industry, but the chairmanship of the Mayor's Committee on Economic Growth, and his membership on the Michigan Economic Development Commission reinforced the knowledge that such planning was not enough. Repeated conversations with Doxiadis in the decade of the sixties made the point explicit. Planning an electrical system for a community, itself without plan or direction, was self defeating.

In 1964, there was public announcement of a joint research program, to begin in 1965 and to extend over five years. In scope, it was probably unprecedented as a private venture.[12] The program was to combine the efforts of three agencies. Doxiadis Associates, from Athens and from the Washington office, would provide the technical services and supervise the entire project. Detroit Edison would support the project both financially and with skilled men, both for the permanent staff and for special assignments. Wayne State University would make available the persons on its faculties whose community knowledge and research skills were pertinent. The University's original commitment to the project was made by President Clarence B. Hilberry; James P. McCormick, then secretary to the University's board of governors, was his representative.

The full name of the ambitious project was "The Developing Urban Detroit Area Research Project," conveniently shortened to UDA. The study was from the first projected to three basic reporting volumes. The first printed volume appeared over the signatures of Walker Lee Cisler, William R. Keast and Constantinos A. Doxiadis, each speaking for his respective institution. The overall title was *The Emergence and Growth of an Urban Region.*[13]

The area under intensive study was defined by the economic relationships which overran political lines. The study included within the Detroit sphere of influence twenty-five counties in Michigan, nine in northern Ohio, and three in Ontario, an area totalling 23,059 square miles and with a population in the 1960 census of 7.1 million people. Wayne County, which contains Detroit, together with Monroe, Washtenaw, Macomb and St. Clair counties had in the 1960 census 52.8 percent of the people of Michigan and 58 percent of the people living in UDA.

The first volume was a study in four scales, specifically the North American Setting, the Great Lakes Megalopolis, the Great Lakes area, and the Urban Detroit Area. The purpose was, with decreasing focus, to inventory and analyze existing conditions, and thus establish the basis on which further projections could rest. The subtitle of volume one was *Analysis.*

The second phase of the study, reported in volume two, (1967) bore the subtitle *Future Alternatives.* It included proposed methodology, and its application to the Urban Detroit Area. Fundamentally, that methodology consisted of the establishment and evaluation of alternatives, and the choice of the preferred one. It was a procedure which, hopefully, could be applied to any urban area.

Changes occurred in scope and focus of the continuing study. The study of the Great Lakes Megalopolis was extracted and given a separate entity. Dr. Harlan Hatcher, member of the board of Detroit Edison and former president of the University of Michigan, accepted the leadership. A non profit corporation would secure support from the business communities extending from Buffalo to Milwaukee; the study would be under the technical direction of Doxiadis Associates, with assistance from the related business and academic communities.

In somewhat the same way, using again the non profit corporation, local application began along the St. Clair River. Business and public leaders from three of the counties included in the larger study—Macomb and St. Clair, in Michigan, and Lambton, in Ontario—formed the International Council. Richard C. Andreae, a member of Edison's board and a business man from Port Huron, led the local organization.

The third volume of the Doxiadis study was scheduled for the summer of 1970, but while it was still in preliminary stages, events occurred which modified the final report. Of these, the most important was the Detroit riot of 1967, and the reaction of the Detroit community to the problems thus thrust forward.

The riot in Detroit in the summer of 1967 was a bundle of contradictions. Perhaps the chief anomaly lies in the fact that this most massive of all urban disruptions occurred in the city which among all the cities of America had the best over all record in employment and housing.

There were slums in the inner city, which had so far escaped the bulldozer, but the riots were not there: they were in the better areas on the fringes of the expanding inner city. There was unemployment in the city, but the majority of those arrested in the riots had jobs, and blacks injured in confrontation with police or militia frequently claimed benefit at hospital under a master insurance program from a place of employment. It was not even a race riot, in the traditional meaning of the term which signified a confrontation of blacks against whites. This was a breakdown of the system of order, accompanied by looting and arson.

The riot first broke out on the margins of one of the distinguished older residential areas of the city. It was, in 1967, a racially mixed neighborhood of well maintained homes. Twelfth Street, which acquired an unhappy fame, and the streets which crossed it, had been, until the current decade, the apartment house and residential area for what could be referred to as the lower middle class. The predominance of black occupants in the area was of relatively recent period, but in both the residential and the business properties on Twelfth Street there was a small but substantial black proprietorship. The buildings were solid, well constructed and in every way modern, though there were some obvious instances of tenant neglect.

From so unlikely a spot, sparked by a minor routine police action which every weekend could match several times over, a riot developed which shocked Detroit and horrified the nation with what seemed to be a racially motivated civil war. Forty-seven persons died violently, and the property damage carried an early estimate of fifty million dollars. As much as twenty-five square miles of the central city were directly involved in the riot. When order was reestablished some five days later, military vehicles and uniformed soldiers gave the manner and dress of a conquered city.

The Detroit Edison Company was directly involved only to a slight degree. Apparently in no place was it directly the object of attack. The fires which spread so widely from so many centers burned poles and transformers, and grounded transmission lines. Some of the Company offices in the affected residential areas were ordered closed, and repair crews were removed until danger of sniper fire could be eliminated.[14] The Company estimated a direct loss, including cost of the restoration of service at about one quarter of a million dollars. Customer accounts which were ordered closed, presumably because of the destruction of the residence, numbered 945, significantly less than the number closed by bulldozer action in each of the immediately preceding years. The indirect loss came from the shock to the business activity, which shows sharply in the 1967 statistics for energy demand and corporate profits.[15]

The reaction of the leaders of the business community to the disorder of 1967 was prompt. Predictably there would be emphasis on law and order. Far more common was the acceptance of collective responsibility for a business society in which this could happen, and a search for remedies was prompt and quite sophisticated.

The New Detroit Committee, a volunteer organization composed of Detroit economic leaders, sat with representatives of community groups and gave attention to both protests and proposals. Walker Cisler was a member of the committee. With the backing of the Ford Foundation and with the resources represented in this business membership, the questions weighed were not whether a thing could be done, but whether it ought to be done.

Beyond question, there was confusion in the counsel. The white community had too often seen the black community as a monolith, and this committee first suffered from the assumption that there must be somewhere a voice to express the needs of the minority as a whole. Instead, it heard a troublesome and sometimes baffling cacophony. There was no simplistic answer to the question of what should be done; the common denominator was the insistence that something be done, and that it be something new.

Fortunately, there was overlap in the membership of the New Detroit Committee with the relatively new MDCDA, and duplication of effort was avoided. The nature of the New Detroit Committee, in any case, tended to minimize tasks which it accepted in its own name, while it threw financial and other support to organizations engaged in approved activities. MDCDA received extensive aid for its work from the New Detroit Committee.

Second only to housing in the popular analysis of the problem was the matter of job availability, and specifically job training. Equally tragic was the number of young blacks, drop outs or high school graduates, who could not or at least did not find jobs. There was also a group who were not hired for physical reasons. Collectively, these minorities within the minority groups were called the "hard core" unemployables, though not all the "hard core" were blacks.

This seemed to be the kind of problem which the business community was especially equipped to deal with. In 1968, a matter of months after the Detroit riots, Henry Ford, II became the articulate leader of a national organization called the National Alliance of Business Men. This was a volunteer group whose purpose, with government assistance, was to rally private businesses to give the required training to the hard core in an effort to include these people in the normal stream of American economic and social life.

On the local Detroit scene, William M. Day became chairman in March, 1968; Day was at that time president of the Michigan Bell Telephone Company. After two years, local leadership went to Edwin O. George, president of Detroit Edison, who received a letter of congratulation and appreciation from the president of the United States. George named Donald J. Pizzimenti, of Detroit Edison, as full time Metropolitan director of NAB for the counties of Wayne, Oakland, and Macomb. At the time of Day's retirement from this leadership, job training had been given to 48,000 disadvantaged persons in the Metropolitan Detroit area, of whom forty-five percent were retained by the training organization.[16]

So far these efforts—nationally, locally, and even within the

Company itself—have achieved only partial success. Although the fact that a number of these who accepted training moved on to permanent employment showed that they had achieved at least a minimal competence, or showed promise of such achievement.[17]

The Doxiadis study of the Developing Urban Detroit Area was directly affected by the urban crisis of 1967. The somewhat philosophical abstractions and generalizations of the first two volumes seemed hardly relevant to a riot scarred city. If the study was to have maximum and continuing impact, the focus must pinpoint the causes of the city's problems, and suggest methods of dealing with them.

Edwin O. George described the situation directly and forcefully. He had been closely involved in the whole Doxiadis study, had been the Company's ranking liaison man, and had approved both of the method and of its results. He emphasized the urgency of the situation by saying, "The problems are at our own back door."

What Ed George meant by that phrase could be made painfully obvious from any upper window in Edison's office buildings. Antiquated and often substandard buildings still stood, but the conspicuous and disheartening aspect was the scores of acres of land which had been reduced to weeds and rubble. The bulldozer had eliminated more low cost housing than had been constructed. The massive efforts to solve the technological problems involved, and to apply the techniques of mass production to housing, had achieved very limited success.

Doxiadis continued to insist that the problem of the inner city was not race but income; that the solution would be found not in housing, be it ever so cheap, but in jobs. Some of those stretching acres of cleared land should be returned to new high labor content industries. He described the ideal situation as one in which a worker could walk from his residence to his job in ten minutes. It was an ideal not likely to appeal to the automobile capital of the world, but it emphasized a possible solution.

By implication, the Doxiadis proposals were a criticism of the thinking which for a quarter of a century had dominated American cities. Zoning had restricted the uses of the land, and thus had effectively limited the expansion of active industries. Some of these had left the state, but more had merely moved from the center of the city to the industrial suburbs or to surrounding towns where space was available.

The impact of this upon the job situation was obvious. The movement of the wealthy and the well to do to the suburb had occurred first, but within the recent two decades industry had followed them, and with industry went those who held jobs. Persons who lived in the inner city found it increasingly difficult to reach the new places of employment, and when they were so employed they too tended to move. There developed a kind of adverse selection process which operated against the inner city and generated further troubles.

The slum, at which men had worked so long, appeared to be the result rather than the cause. Deserted by those able to get out, its population reduced in numbers and still more in wealth, and scarred by demolition and abandonment, the area erupted with every social disorder such a situation can produce.

To build mass housing in this area of low achievement would seem to compound its ills. Dick Hart, who had represented the Company in many of these frustrating conferences, also became convinced that housing must follow jobs. He felt that considerations of job availability and population mix alone would make any kind of mass housing socially acceptable. Otherwise there was real danger that Detroit would merely replace its old with newly built slums.

The 1967 crisis underlined the importance of the Doxiadis approach. That conception was that it was essential to the whole city that life be restored to the core area. It must again become the creating and integrative center of urban activity and culture. To return life to that part of the metropolitan complex, meant bringing in people, and with them the services they would require there. Doxiadis was perhaps underlining the obvious in this report, but the logic was complete. Those acres of land from which slums had been cleared must at least in part be returned, not to the dispossessed former resident of these substandard structures, but to the higher income group who by their presence there could revitalize the area, and with it the metropolitan complex.

There was no natural solution to the problems of the city or the inner city; left to itself, the spreading area of social and physical decay would find no dependable boundary. The Doxiadis report argued this, and subsequent developments seemed to confirm the fact; massive and positive efforts were necessary.

This kind of development cannot be successfully accomplished by isolated individual efforts or in fragmentary planning. The Detroit Edison Company proposed to interest other corporate entities in the problem, and in the meantime set a pattern with its own program. Edison's two big office buildings were located in a clutter of parking lots and antiquated structures. Working with Doxiadis and with Detroit city planners, Edison evolved a two phase program for the development of this and the surrounding area, to be called Edison Center.

Fifty-two acres were accumulated in this area, defined by major streets and expressways. The homes which had existed there had been modest, turn-of-the-century types. Fewer than fifty residential structures still remained.

Phase one called for the immediate completion of a twenty-five story office building, a parking structure and an elevated, three acre plaza development to tie the new building to Edison's existing buildings. Detroit Edison will not build or finance that building, but will lease several floors for Company use. The cost of the structure was estimated by the builders at eighteen million dollars.

A long range program, phase two, is under consideration for the remainder of the fifty-two acre area. It anticipates a complex of high rise and low profile housing for several income levels, which Edison employes, and many other down town workers will, hopefully, find convenient. A land use study and a marketing feasibility study by consulting firms recommended the kind of housing, shops and commercial buildings which appear to be most desirable and practical.[18]

The development is not intended to be a philanthropy. It is, however, a self-conscious action on the part of Detroit Edison to reinvigorate the central business district. Edison has made a commitment, in expectation that others of the business community will find it possible to undertake similar action.

CHAPTER EIGHTEEN

TOWARD AN ORDERED WORLD

The Acceptance of Responsibilities / Postwar Reconstruction and Assistance / The Businessman's Search for Understanding / The World Energy Conference; Moscow and Detroit.

THE DETROIT EDISON COMPANY'S FRANCHISED SERVICE WAS LIMITED TO southeast Michigan, but the Company looked beyond its borders for those advances in power technology which would aid in the discharge of this responsibility. The nation had made remarkable progress in the field of energy, and the Company was proud of its participation in that development.

It was in fact an international achievement. Alex Dow was proudly British, and he maintained both social and professional associations with Britain and the Continent. He made regular visits there, and by personal observation kept the Company abreast of the activities in these wider fields. His associates and his immediate successors shared this determination.

These tentative first steps became a broad avenue of trend in the decades which followed. There is in this of course an historical development which can and will be traced. It is obvious that the changing sense of time and space has brought distant continents closer and with it a sense of involvement. It is possible thus to identify an original impulse within the Company, but the directed continuation and extension of this concern with international matters stems from Walker Lee Cisler. He accepted as a personal moral responsibility the obligation to do something about this troubled world, and in this resolution he carried along with him the board of directors and many of the staff.

The story then, of Detroit Edison as a corporate citizen, with the citizen's concern for international welfare and world order, must become also the story of Walker Cisler's personal devoted service to this cause. On one occasion, he made the offhand estimate that he was spending one third of his waking hours on international affairs. Cisler found a deep satisfaction in his leadership of a great and powerful corporation, and in the role which allowed him to direct its activities toward what he believed to be desirable ends. A colleague, who knew him well, said that this reward was Walker Cisler's real compensation for his work for The Detroit Edison Company.

As an officer with the Allied forces in Europe, Cisler had had the responsible task of returning to service many of the utilities which the war had destroyed. In this service, and the advisory work with the postwar governments, he was made acutely conscious of the waste and the attendant human suffering. A number of times in later years, Cisler referred to this as a shaking experience, and to his resolution then to make energy a creative fact, contributing to man's well-being as dramatically as it had contributed here to the devastation of man's works.

Because of his unique knowledge of European electric systems and his personal acquaintance with leaders in industry and government in each nation, he was asked to assist in developing the electric power aspects of the Marshall Plan in 1948. Although he was then anxious to establish himself in Detroit Edison affairs and the Detroit Edison business community, he accepted the responsibility as a duty.

The European complement to the Marshall Plan was the seventeen-nation Organization for European Economic Cooperation, formed some two weeks after Congress passed the Foreign Aid Assistance Act. Initially serving as the planning and distributing agency for American assistance, OEEC later found a more permanent position as the high level body for the improvement of economic conditions among the member nations. It inspired and helped to bring about the European Coal and Steel Community, the Common Market and Euratom. Cisler became the American representative on OEEC.

In 1956 OEEC published a landmark report entitled *Europe's Growing Needs of Energy—How Can They Be Met?* and in 1960 an even more comprehensive report, *Towards a New Energy Pattern in Europe.* Engineers and economists worked side by side on these reports in forecasting energy requirements of all kinds for European economic development through 1975, testing the projections against the capital that would be available and that required. An interesting result was the finding that about twelve percent of all new investment would be needed to make energy available and that seven times as much would be needed for all the facilities to utilize the energy. Since then much the same ratio has been found to apply to economic growth all over the world.

Earlier experience in the United States provided the basis for three types of studies. The first and simplest of these was the Statistical Electric Power Survey, modeled along the lines of the Semiannual Electric Power Surveys for the United States, first suggested by Cisler and made regularly after 1947 by the Edison Electric Institute. Once the first survey was actually completed, it was possible for the local people to make it regularly. Nations as far apart geographically and economically as Japan and Turkey have found such regular, accurate knowledge of electric generating capacity, future load growth and heavy electric equipment needs and construction capacity invaluable.

A somewhat more extensive study was desirable in developing countries which had limited experience in business methods. This was the analytical electric power survey, which again was made on a shoulder-to-shoulder basis with the nationals, beginning with an inventory of electric facilities, projections as to future load requirements, economic opportunities, desirable system interconnections, standards and developments, rate structure, financial requirements, and the training needs of technical and management personnel.

The third form of technical assistance was the total energy study, which concerned itself with energy in all forms, much after the pattern of the OEEC studies. The first of these was made in

Tunisia in 1961 and a voluminous two-volume study, requiring some two years to complete, was finished in 1963 for India, where the government did it the signal honor of accepting it as an official document.[1]

The influence of one Company, and the extent of its involvement, can best be indicated by a listing, certainly an incomplete one, of the countries in which Edison men were called to work on technical assistance. These included the successive programs of postwar recovery, economic cooperation, aid to undeveloped countries, and ultimately assistance to Central and South America. There are twenty-one names on the list: Argentina, British Guinea, Canada, Great Britain, France, Greece, India, Iran, Italy, Japan, Korea, Madagascar, Nigeria, Okinawa, the Philippines, Vietnam, Taiwan, Thailand, Tunisia, Turkey, and Venezuela.[2]

The flow of personnel was not all one way. Technical assistance included training men to use equipment, or to modify methods, and it proved convenient to bring some of them to this country. The training program ranged all the way from a mere tourist's quick look at a machine to a serious assignment which could last for months. The first of these foreign visitors to Detroit Edison was welcomed in 1949 and by 1962 the 1,000th group had come from sixty-two countries.[3]

The numerous visitors to the Company from abroad are difficult to classify. Some, but certainly a minority, came at the expense of one or another of the aid programs of the United States. Some represented the aspirations of the employing company, which covered the visitor's expense. In some cases, the visitors were government sponsored and supported, especially in areas where the government operated comparable facilities. For such visitors the out of pocket direct cost to the Company was relatively small.

For Detroit Edison, there has been one uncalculated gain from these activities, especially from those involving service abroad. As management well knew, it is one thing to be an executive or an engineer in a successful electric utility which can provide confident support; it is quite another thing to work and to succeed in utility matters in a developing country where there is no such support. A significant number of Detroit Edison's key personnel had this experience, and the Company is stronger as a consequence.

The assistance programs of the postwar period were intended to reestablish a viable economy in the wartorn areas of Europe and Asia. With the cooperation of the United States government, and with its financial support, that aid was extended to the underdeveloped nations. The interest of the United States government in the establishment and maintenance of an orderly world justified its actions.

It is impossible to say when businessmen became conscious that they too had an interest in recreating a world of order, and that their normal business relationships constituted a significant medium for the advancement of international understanding. Completely outside the formal channels of diplomacy, businessmen could express common hopes and fears and discover common interests, and in so doing they tended to reduce international confrontations to the human scale, and to the personal level.

In this normal business world, Edison could be a part. Foreign bids were invited on a number of items, and cost comparisons on occasion justified an international purchase. Thus, some special forged steam pipe, for high pressure use, came economically from Japan. A small turbine generator in the Beacon Street heating plant came from Sweden, and was thus a precedent for the big machine from the same country placed at Harbor Beach.[4] There were other items, relatively small in their total, in which foreign manufacturers met or bettered American prices.[5]

With the relaxation of the cold war, there was hope that this kind of personal association could be extended to eastern Europe. Here it would take the form of technical cooperation extended to the Soviet Union and some of the nations within its sphere of influence.

Cisler made his first visit to the Soviet Union in 1958 and went again in 1959, leading groups of utility officials sponsored by the Edison Electric Institute under technical exchange agreements between the State Department and the Soviets. These visitations were reciprocated by Soviet scientists and engineers who shared the interest of the Americans in advances in electric power generation and transmission, and in the new field of nuclear power for peaceful purposes.[6]

The atmosphere on these exchanges was cordial and nonpolitical. Despite the language barrier, they shared the common language of the kilowatt, which does the same work whether it is applied in the Soviet Union or in the United States. The language of the engineer is international and it helps to bring nations together for peaceful and constructive purposes.

Cisler was an immediate success with the engineers he met in the Soviet Union because they responded to his love for power plants, imaginative transmission projects and large-scale industrial developments. His robust ability to keep to a hard schedule of plant inspections on very little sleep, in all sorts of weather, won their admiration. Like many of them, he came from a farming background, knew about soil, crops and seasons just as he understood sophisticated technical matters.

Over the years from 1959 onward Cisler led technical exchange visitations at the invitation of the respective governments to Poland, Romania, Czechoslovakia and Bulgaria, which in turn led to similar visits to the United States.

Others from Detroit Edison over the years have participated in technical exchanges and received excellent cooperation, with access to power plants across the Soviet Union. On one such visit Harvey Wagner, then executive vice president for engineering, observed a slogan over the entrance to a nuclear plant, which expressed the hope of many in these person-to-person exchanges: "May the atom be a worker and not a soldier."

The earlier and deliberate effort to stimulate world trade as a means of international understanding continued in Edison. Given the economic situation, there was a price factor which also justified the purchases of special equipment abroad. Three contracts, all in 1969, illustrate the international character of the technology, the economy of such transactions, and the deliberate desire to advance international goodwill through commercial contacts. The effort to include the Soviet Union in this circle of business friendships seemed to be logical and desirable.

From Britain, the purchase contract called for delivery of the 1,150,000 kw generator for installation at Fermi II. From Japan, the two companies which were building the pumped storage facility at Ludington purchased the complicated pumping-generating units. From the Soviet Union, where a respected competence in high voltage transmission had been demonstrated, the Company ordered a 345,000 volt transformer for the new ehv system.

In Moscow 1968 Walker Cisler, newly elected Chairman of the International Executive Council (governing body of the 65-nation World Energy Conference), shakes hands with the newly elected President of the WEC, Pyotr S. Neporozhny, USSR Minister of Energy and Power.

The task of writing the contract, establishing price and negotiating marketing terms with a branch of the Soviet government proved to be difficult. Even delivery in commercial international exchange was strange to the government-owned producer and the solution to these problems gave Edison men some new experiences. The negotiations over the transformer required several persons to visit Russia and several from that side had to visit the United States.

Among the many honors which came to Walker Cisler over the years was one which in 1968 brought special distinction, the election to chairmanship of the International Executive Council of the World Energy Conference. This was a distinguished international organization of some sixty-eight nations, of which the traditional center was London. Cisler was the first chairman who was not British.

The organization was begun in 1924 when the leading industrial nations sent representatives to the first World Power Conference. The American delegation there presented the resolution which on adoption made the organization permanent. Thereafter, except for the war's interruption, the World Power Conference met in plenary session every six years, with some additional sectional meetings in intervening years. The International Executive Council, which was the executive body, met annually.

It is traditional in the Conference that the presidency go to a representative of the host country. At the 1968 conference in Moscow at which Cisler received the executive responsibility, the honor was given to Pyotr Neporozhny, Minister for Power and Electrification of the Soviet Union. His term as president would expire in 1971, when the plenary conference was scheduled for Bucharest.

When the Conference met in Haifa in 1965, a review committee was named to report at Moscow. Under the chairmanship of Erik Grafstrom, president of the Swedish State Power Board, the membership included representatives from Australia, France, Great Britain, Japan, Nigeria, Poland, the United States, and the Soviet Union. Cisler was the United States' representative. The purpose of the committee was to recommend changes which should be made in the organization in the light of contemporary circumstances.

At Moscow, the committee report led to the adoption of several proposals. Plenary sessions were scheduled at three rather than at six year intervals, though the term of the Chairman of the Executive Committee remained at six years. Provision was made for special study committees, on problems of common interest, on a continuing basis. A new name, World Energy Conference, described more accurately the larger fields of interest, and in the rewritten constitution there was an added item in the statement of the organization's purposes. It was resolved that the Conference be concerned with "energy consumption in its overall relationship with the growth of economic activity in the area."[7]

This is a new note in the affairs of the World Energy Conference. The meetings and the activities of the Conference had been rigorously technical. The new statement of objectives was mild and very general. It did, however, seem to open the way for a range of activities especially in the developing world which would reflect a sense of moral obligation. Much would depend on what the leaders did with that statement.

Walker Cisler, as chairman of the International Executive Committee, could hope to accomplish two specific objectives. Personal friendships and individual confidence could be strengthened—leading to a significant decrease in international tension. Secondly, that cautious statement of purpose which was adopted could make the World Energy Conference an instrument, not just for accumulation and distribution of technical information, but for the application of technology for human good, especially in the underdeveloped areas.[8]

Responsibility for the leadership in this organization and for the development of its program became a personal thing for the chairman. He announced his determination to visit as many of the sixty-eight member states as possible. Since he had visited fifty-two nations in the quarter century after World War II, usually on official mission, the proposed trips would allow the renewal of old friendships and associations.

At a 1966 meeting in Japan, Cisler, as chairman of the U. S. National Committee, presented an invitation to the International Executive Committee to hold the 1974 World Energy Conference in Detroit. This would be the first such meeting in the United States since just before World War II. The subsequent election of Cisler as chairman gave him general responsibility for both the program and the organization of that Detroit Conference.[9]

This will be the silver anniversary of the World Energy Conference, giving the meeting a special distinction. Delegates and guests from some sixty-eight member nations will number above six thousand persons. The meeting will also coincide with the completion of Cisler's six year term as chairman of the International Executive Committee. A World Energy Conference in Detroit in 1974 will provide a unique and appropriate landmark in a long and distinguished career.

Appendix A

GLOSSARY OF ABBREVIATIONS

AEC	Atomic Energy Commission
APDA	Atomic Power Development Associates
Btu	British thermal unit
CANUSE	Canadian U.S. Eastern
C and M	Construction and Maintainence
Company	Detroit Edison Company, The
ECAR	East Central Area Reliability Coordination
EEI	Edison Electric Institute
ehv	extra high voltage
FPC	Federal Power Commission
hv	high voltage
Hydro	Ontario Hydro-Electric Commission
kva	kilovolt ampheres
kw	kilowatt
kwhr	kilowatt hour
MDCDA	Metropolitan Detroit Citizens Development Authority
MIIO	Michigan, Illinois, Indiana, Ohio
mw	Megawatt
NERC	National Electric Releability Council
PJM	Pennsylvania, New Jersey, Maryland
PLC	Public Lighting Commission
PRDC	Power Reactor Development Company
ps	pumped storage
REA	Rural Electrification Administration
SEC	Securities and Exchange Commission
TVA	Tennessee Valley Authority
UDA	The Developing Urban Detroit Area Research Project
uhv	ultra high voltage

Appendix B

FINANCIAL AND OPERATING STATISTICS OF THE DETROIT EDISON COMPANY

CONTENTS:

(Thousands of Dollars)

Year (At December 31)	Assets: Utility Properties	Assets: Current	Assets: Miscellaneous	Assets: Total	Liabilities: Stock-holders Equity	Liabilities: Long Term Debt	Liabilities: Current	Liabilities: Depreciation and other Reserves	Liabilities: Other	Liabilities: Total
1942 (Restated)	347,288	28,174	5,407	380,869	160,742	138,310	20,183	61,579	55	380,869
1947	383,549	42,507	8,010	434,066	169,223	145,000	22,141	97,586	116	434,066
1952	623,608	54,261	1,727	679,596	254,163	264,153	38,790	121,745	745	679,596
1953	695,429	63,563	1,713	760,705	270,011	313,785	44,011	132,252	646	760,705
1954	761,276	56,515	1,886	819,677	276,615	349,901	45,386	147,775	-	819,677
1955	850,390	58,506	2,703	911,599	285,368	405,133	54,159	166,939	-	911,599
1956	929,474	66,910	2,208	998,592	331,038	418,558	60,833	188,163	-	998,592
1957	1,000,397	87,200	2,800	1,090,397	339,883	483,988	57,137	209,389	-	1,090,397
1958	1,065,720	59,437	2,694	1,127,851	376,378	437,553	79,922	233,998	-	1,127,851
1959	1,116,023	70,344	3,848	1,190,215	393,421	453,293	85,107	258,394	-	1,190,215
1960	1,168,692	64,164	3,957	1,236,813	404,512	450,841	98,241	283,219	-	1,236,813
1961	1,195,193	65,396	5,177	1,265,766	413,948	449,106	94,189	308,523	-	1,265,766
1962	1,220,743	73,501	8,578	1,302,822	425,183	447,927	92,809	336,903	-	1,302,822
1963	1,252,882	78,606	7,198	1,338,686	436,985	440,968	95,119	365,614	-	1,338,686
1964	1,280,431	91,057	8,010	1,379,498	453,602	440,064	90,647	395,185	-	1,379,498
1965	1,333,715	92,754	7,918	1,434,387	473,165	439,376	96,770	425,076	-	1,434,387
1966	1,444,917	134,226	8,954	1,588,097	494,671	538,769	97,855	456,802	-	1,588,097
1967	1,619,745	123,961	10,257	1,753,963	607,818	538,160	119,898	488,087	-	1,753,963
1968	1,772,142	155,898	21,305	1,949,345	622,384	679,561	123,242	524,158	-	1,949,345
1969	1,974,199	114,043	25,915	2,114,157	635,320	761,793	145,192	571,852	-	2,114,157

CONSOLIDATED INCOME STATEMENT AND RETAINED INCOME STATEMENT DATA

(Thousands of Dollars)

Year	Total Revenues	Total Expenses	Interest and Other Deductions from Income	Net Income	Retained Income: Balance 1st of Year	Dividends	Other Charges	Balance Year End
1945	85,817	68,682	9,204	7,931	33,143	7,633	380	33,061
1946	90,917	75,860	4,426	10,631	33,061	7,634	-265	36,323
1947	106,961	90,371	5,477	11,113	36,323	8,204	13,901	25,331
1948	120,223	105,041	4,731	10,451	25,331	8,395	-261	27,648
1949	134,632	115,188	6,134	13,310	27,648	8,601	-340	32,697
1950	150,812	126,145	7,635	17,032	32,697	9,244	863	39,622
1951	164,317	140,260	8,637	15,420	39,622	10,938	-3,925	48,029
1952	173,419	148,348	7,875	17,196	48,029	13,564	None	51,661
1953	192,570	163,301	8,749	20,520	51,661	16,244	None	55,937
1954	196,000	162,794	11,069	22,137	55,937	17,215	849	60,010
1955	219,712	182,032	11,383	26,297	60,010	17,812	None	68,495
1956	236,685	195,467	11,841	29,376	68,495	22,254	2,362	73,256
1957	251,026	207,550	10,758	32,718	73,256	24,943	None	81,031
1958	246,504	203,847	12,556	30,101	81,031	25,544	10,560	75,029
1959	267,287	220,375	13,482	33,430	75,029	28,297	-	80,162
1960	279,302	227,063	13,781	38,457	80,162	29,308	-	89,312
1961	285,798	231,110	14,923	39,765	89,312	31,556	-	97,520
1962	301,446	242,908	15,626	42,911	97,520	32,352	-	108,080
1963	320,034	258,033	15,827	46,174	108,080	34,545	193	119,516
1964	343,620	276,511	15,385	51,724	119,516	35,283	-	135,956
1965	364,684	291,535	15,585	57,564	135,956	38,181	-	155,340
1966	393,959	317,889	14,039	61,981	155,340	40,354	228	176,739
1967	413,232	337,732	14,875	60,624	176,739	41,594	-	195,770
1968	444,136	367,800	16,309	60,027	195,770	45,659	-	210,138
1969	483,183	396,914	24,841	61,428	210,138	45,672	2,868	223,026

REVENUES BY CLASSES OF SERVICE

(Thousands of Dollars)

Year	Electric				Steam		
	Domestic	Commercial	Industrial	Misc.	Heating	Other	Total
1945	28,750	21,508	28,020	3,834	2,510	1,195	85,817
1946	31,230	23,606	28,234	4,706	2,383	758	90,917
1947	34,988	28,496	33,540	6,174	2,819	944	106,961
1948	39,066	32,233	38,136	6,099	3,401	1,288	120,223
1949	46,995	38,383	38,826	5,498	3,855	1,075	134,632
1950	52,223	41,971	44,966	6,161	4,232	1,259	150,812
1951	58,241	45,363	49,230	6,608	4,352	523	164,317
1952	63,398	47,725	49,910	6,923	4,568	895	173,419
1953	69,225	51,806	58,095	8,075	4,817	552	192,570
1954	75,247	53,469	53,524	8,450	5,146	164	96,000
1955	80,772	58,242	65,990	9,231	5,241	236	.19,712
1956	88,851	63,026	68,276	10,080	6,442	10	236,685
1957	94,334	65,784	72,699	10,741	7,026	442	251,026
1958	97,564	65,490	65,036	10,892	7,181	341	246,504
1959	103,517	72,018	71,764	12,165	7,566	257	267,287
1960	107,632	74,652	75,761	13,790	7,115	352	279,302
1961	112,406	76,477	75,055	14,456	7,317	87	285,798
1962	115,932	80,993	81,840	15,231	7,334	116	301,446
1963	120,850	84,855	89,492	16,894	7,387	556	320,034
1964	128,450	90,639	97,511	19,034	7,161	825	343,620
1965	134,875	93,203	106,950	20,821	7,419	1,416	364,684
1966	145,378	99,422	113,113	27,926	7,581	539	393,959
1967	155,032	103,088	117,959	27,893	7,765	1,495	413,232
1968	166,589	108,722	129,421	30,819	7,651	934	444,136
1969	178,784	116,387	142,166	36,501	8,166	1,179	483,183

OPERATING EXPENSES BY PRINCIPAL ITEMS
(All Dollars in Thousands)

Year	Payrolls and Pensions	Fuel	Other Operating and Maintenance Expenses	Retirement (Depreciation) Appropriations	Taxes	All Other	Total Expenses	Maintenance & Depreciation as % of Utility Revenues	Taxes as % of Total Revenues
1945	24,098	15,098	8,090	9,330	11,938	128	68,682	18.6	13.9
1946	28,576	16,422	9,252	7,620	13,712	278	75,860	16.5	15.1
1947	32,207	23,879	13,110	7,620	13,126	429	90,371	16.4	12.3
1948	35,126	30,590	15,704	8,220	14,972	429	105,041	16.0	12.5
1949	38,762	30,285	17,573	9,120	19,019	429	115,188	15.7	14.1
1950	42,340	33,320	18,445	10,110	21,501	429	126,145	15.0	14.3
1951	47,440	35,339	21,022	11,400	24,584	475	140,260	15.6	15.0
1952	51,162	36,176	21,111	13,200	26,273	426	148,348	16.1	15.2
1953	52,683	41,340	23,514	14,700	30,638	426	163,301	16.0	15.9
1954	54,181	35,901	23,613	16,230	32,443	426	162,794	16.6	16.6
1955	56,920	38,910	27,348	18,000	40,428	426	182,032	16.3	18.4
1956	59,399	43,459	27,603	19,900	45,106	-	195,467	15.6	19.1
1957	60,836	47,866	29,603	21,900	47,345	-	207,550	15.1	18.9
1958	62,137	44,060	30,483	23,640	43,527	-	203,847	15.8	17.7
1959	63,482	46,102	31,881	25,560	53,350	-	220,375	15.7	20.0
1960	64,872	46,847	33,647	27,108	54,589	-	227,063	15.7	19.5
1961	65,795	46,260	33,898	27,804	57,353	-	231,110	16.4	20.1
1962	66,519	51,827	33,407	29,484	61,671	-	242,908	16.5	20.5
1963	68,237	56,847	33,276	31,314	68,359	-	258,033	16.5	21.4
1964	71,957	61,601	37,986	31,885	73,082	-	276,511	16.7	21.3
1965	74,187	67,057	44,825	32,919	72,547	-	291,535	16.5	19.9
1966	82,548	80,666	52,212	34,441	68,022	-	317,889	16.5	17.3
1967	86,948	86,220	61,112	36,735	66,717	-	337,732	17.9	16.1
1968	93,677	95,838	55,450	40,090	82,745	-	367,800	17.8	18.6
1969	93,488	112,149	57,100	46,188	87,969	-	396,914	17.6	18.2

CUSTOMERS BY CLASSES OF SERVICE

(At Year End)

Year	Electric					Steam	
	Domestic	Commercial	Industrial	Misc.	Total	Heating	Other
1945	726,271	83,154	560	486	810,471	1,963	13,353
1946	744,186	89,230	502	493	834,411	2,037	13,794
1947	771,358	92,976	488	491	865,313	1,937	14,104
1948	802,509	95,860	483	502	899,354	1,889	13,843
1949	831,776	98,128	499	555	930,958	1,755	13,405
1950	871,008	99,795	530	600	971,933	1,771	12,901
1951	911,249	100,780	566	628	1,013,223	1,864	761
1952	936,566	102,146	589	640	1,039,941	1,858	770
1953	974,454	104,295	620	655	1,080,024	1,838	788
1954	1,006,468	106,643	639	682	1,114,432	1,854	781
1955	1,044,964	107,564	664	711	1,153,903	1,814	787
1956	1,080,780	110,090	703	755	1,192,328	1,758	777
1957	1,114,111	111,238	714	779	1,226,842	1,696	783
1958	1,134,145	112,966	713	819	1,248,643	1,629	819
1959	1,154,326	113,459	732	883	1,269,400	1,679	809
1960	1,167,548	113,960	752	920	1,283,180	1,600	801
1961	1,177,364	113,809	769	946	1,292,888	1,527	823
1962	1,189,984	114,427	786	989	1,306,186	1,444	823
1963	1,206,668	115,066	845	1,028	1,323,607	1,393	838
1964	1,235,028	116,612	898	1,079	1,353,617	1,362	841
1965	1,270,950	112,922	986	1,118	1,385,976	1,348	847
1966	1,305,379	113,567	1,084	1,158	1,421,188	1,317	861
1967	1,332,522	113,793	1,186	1,174	1,448,675	1,296	875
1968	1,358,684	114,383	1,302	1,192	1,475,561	1,279	878
1969	1,390,751	115,502	1,433	1,214	1,508,900	1,252	882

KWH SALES BY CLASSES OF SERVICE

(Thousands of KWHS)

Year	Domestic	Commercial	Industrial	Misc.	Total
1945	1,010,305	857,798	2,497,837	274,904	4,640,844
1946	1,111,949	945,935	2,310,809	328,790	4,697,483
1947	1,243,765	1,137,835	2,826,328	466,432	5,674,360
1948	1,374,521	1,268,336	3,084,813	408,681	6,136,351
1949	1,508,793	1,340,691	3,058,130	359,918	6,267,532
1950	1,695,725	1,467,774	3,721,412	424,272	7,309,183
1951	1,935,140	1,607,396	4,102,627	414,792	8,059,955
1952	2,142,906	1,702,416	4,116,763	436,768	8,398,853
1953	2,381,161	1,869,341	4,906,181	508,795	9,665,478
1954	2,632,067	1,932,600	4,445,856	515,191	9,525,714
1955	2,859,629	2,165,076	5,821,551	591,909	11,438,165
1956	3,179,075	2,328,517	5,822,006	529,256	11,858,854
1957	3,426,065	2,456,772	6,111,156	565,244	12,559,237
1958	3,580,998	2,440,424	5,235,593	516,489	11,773,504
1959	3,850,226	2,715,599	5,957,348	633,556	13,156,729
1960	4,032,843	2,829,102	6,423,652	719,959	14,005,556
1961	4,243,054	2,898,632	6,308,959	721,789	14,172,434
1962	4,395,232	3,102,985	7,106,771	829,390	15,434,378
1963	4,607,999	3,283,056	7,969,661	945,673	16,806,389
1964	4,938,211	3,536,797	8,892,029	1,022,909	18,389,946
1965	5,335,118	3,733,248	9,981,413	1,156,654	20,206,433
1966	5,852,180	4,017,797	11,200,554	1,256,259	22,326,790
1967	6,323,198	4,192,971	11,395,599	1,266,937	23,178,705
1968	6,990,044	4,485,764	12,605,066	1,513,855	25,594,729
1969	7,649,327	4,808,709	13,751,609	1,995,499	28,205,144

PLANT INVESTMENT, RETIREMENT AND DEPRECIATION DATA
(All Dollars in Thousands)

Year	Plant Additions	Plant Retirements	Year End Balance	Depreciable Plant Year End	Appropriations to Retirement Reserve	% of Appropriations to Average Depreciable Plant	Retirement Reserve Year End	% Retirement Reserve to Depreciable Plant
1945	5,541	2,130	354,610	335,139	9,355	2.8	84,792	25.3
1946	13,337	3,019	364,928	345,324	7,646	2.2	90,304	26.2
1947	22,904	3,899	383,933	363,553	7,705	2.2	95,035	26.1
1948	37,448	5,073	416,308	378,705	8,432	2.3	98,737	26.1
1949	51,412	6,823	460,897	423,238	9,449	2.4	102,303	24.2
1950	59,438	10,464	509,871	452,556	10,473	2.4	107,175	23.7
1951	62,054	11,044	560,881	524,579	11,769	2.4	111,617	21.3
1952	70,181	6,813	624,249	549,311	13,509	2.5	119,630	21.8
1953	78,630	6,833	696,046	635,121	15,000	2.5	129,128	20.3
1954	73,637	7,699	761,984	712,591	16,558	2.5	139,660	19.6
1955	97,953	8,611	851,326	750,825	18,312	2.5	151,954	20.2
1956	89,284	10,019	930,591	837,461	20,197	2.5	165,489	19.8
1957	81,830	10,925	1,001,496	904,226	22,316	2.6	178,142	19.7
1958	75,087	9,656	1,066,927	976,779	24,172	2.6	194,033	19.8
1959	59,235	8,609	1,117,554	1,050,746	26,137	2.6	212,419	20.2
1960	61,777	8,936	1,170,395	1,073,117	27,576	2.6	232,038	21.6
1961	36,838	9,749	1,197,485	1,155,149	28,169	2.5	251,663	21.7
1962	36,322	7,557	1,226,250	1,180,701	29,901	2.6	274,131	23.2
1963	41,605	10,877	1,256,977	1,204,362	31,327	2.6	297,130	24.6
1964	36,931	9,207	1,284,702	1,233,856	32,502	2.7	321,430	26.0
1965	63,669	10,003	1,338,368	1,272,058	33,623	2.7	345,992	27.1
1966	123,153	11,284	1,450,237	1,332,387	36,494	2.8	372,162	27.9
1967	189,594	14,036	1,625,795	1,415,625	37,965	2.8	396,017	27.9
1968	171,517	17,775	1,779,537	1,579,202	41,520	2.8	420,302	26.6
1969	221,873	16,936	1,984,474	1,780,434	47,822	2.8	451,651	25.3

SHAREHOLDER DATA
(Common Stock)

Year	Shareholders Year End Number	Shares Outstanding Year End Thousands	Net Income Per Share Dollars	Dividend Declared Per Share Dollars	% Dividend Payout	Price Range Capital Stock High	Price Range Capital Stock Low
						$100 Par	
1945	37,214	6,361	1.25	1.20	96.0	25-1/2	21
1946	38,207	6,361	1.67	1.20	71.9	28	23
1947	40,224	6,995	1.59	1.20	75.5	27-1/2	20-7/8
1948	52,154	6,996	1.49	1.20	80.5	21-3/4	20
1949	53,345	7,677	1.73	1.20	69.4	23-1/4	20
1950	54,958	7,805	2.18	1.20	55.0	24	21-3/8
1951	59,765	9,175	1.68	1.25	74.4	23-5/8	21-5/8
1952	66,184	10,082	1.71	1.40	81.9	25-1/8	23
1953	71,682	10,675	1.92	1.55	80.7	30-7/8	24-1/2
1954	75,516	10,787	2.05	1.60	79.1	35	28-3/8
1955	77,476	10,800	2.43	1.65	67.9	37-3/4	33-1/2
1956	88,198	12,439	2.36	1.85	78.4	38-7/8	33-7/8
1957	93,062	12,483	2.62	2.00	76.3	41-7/8	35-1/4
1958	100,829	13,871	2.17	2.00	92.2	42-3/8	37-7/8
1959	105,936	14,259	2.34	2.00	85.5	47-3/4	41-1/4
1960	107,328	14,323	2.68	2.05	76.5	49-3/8	40-1/8
1961	106,384	14,364	2.77	2.20	79.4	67-3/4	48-3/8
						$10 Par	
1962	106,558	28,772	1.49	1.125	75.5	31-5/8	30
1963	111,324	28,796	1.60	1.20	75.0	35-3/4	31-1/2
1964	113,302	28,809	1.80	1.225	68.1	37-3/4	31-7/8
1965	115,404	28,821	2.00	1.325	66.3	38-5/8	34-7/8
1966	118,401	28,828	2.15	1.40	65.1	37-1/2	28-1/8
1967	120,224	28,835	2.07	1.40	67.6	34-1/4	26
1968	125,435	28,851	1.90	1.40	73.7	29-3/8	24
1969	126,315	28,854	1.95	1.40	71.8	29-5/8	20-1/4

FINANCIAL RATIOS AND CAPITAL STRUCTURE

	Operating Ratios-%			Total Plant per Dollar of Utility Revenues	Capital Structure				
					Thousands of Dollars			Debt-Equity Ratios-%	
Year	All Expense	Excluding Depreciation	Excluding Depreciation & Taxes	Dollars	Long Term Debt	Stock-holders Equity	Total Capital Structure	Debt	Equity
1945	80.0	69.7	55.7	4.16	115,000	164,078	279,078	41.2	58.8
1946	83.4	75.2	60.1	4.02	115,000	167,416	282,416	40.7	59.3
1947	84.5	77.5	65.2	3.60	145,000	169,223	314,223	46.1	53.9
1948	87.4	80.8	68.3	3.47	191,641	171,618	363,259	52.8	47.2
1949	85.6	78.9	64.7	3.43	201,641	190,264	391,905	51.5	48.5
1950	83.6	77.1	62.8	3.39	229,482	199,809	429,291	53.5	46.5
1951	85.4	78.6	63.6	3.42	283,294	232,390	515,684	54.9	45.1
1952	85.5	78.3	63.1	3.62	264,153	254,163	518,316	51.0	49.0
1953	84.8	77.4	61.4	3.62	313,785	270,011	583,796	53.7	46.3
1954	83.1	74.8	58.3	3.89	349,901	276,615	626,516	55.8	44.2
1955	82.9	74.7	56.3	3.88	405,133	285,368	690,501	58.7	41.3
1956	82.6	74.2	55.1	3.93	418,558	331,038	749,596	55.8	44.2
1957	82.7	74.1	55.2	4.00	483,988	339,883	823,871	58.7	41.3
1958	82.7	73.2	55.5	4.33	437,553	376,378	813,931	53.8	46.2
1959	82.4	72.9	53.0	4.18	453,293	393,421	846,714	53.5	46.5
1960	81.3	71.7	52.1	4.19	450,841	404,512	855,353	52.7	47.3
1961	80.9	71.1	51.1	4.19	449,106	413,948	863,054	52.0	48.0
1962	80.6	70.8	50.4	4.07	447,927	425,183	873,110	51.3	48.7
1963	80.8	71.0	49.6	3.93	440,968	436,985	877,953	50.2	49.8
1964	80.6	71.3	50.0	3.75	440,064	453,602	893,666	49.2	50.8
1965	80.5	71.5	51.5	3.68	439,376	473,165	912,541	48.1	51.9
1966	80.8	72.0	54.7	3.69	538,769	494,671	1,033,440	52.1	47.9
1967	82.0	73.1	56.9	3.95	538,160	607,818	1,145,978	47.0	53.0
1968	83.0	73.9	55.3	4.01	679,561	622,384	1,301,945	52.2	47.8
1969	82.3	72.8	54.5	4.12	761,793	635,320	1,397,113	54.5	45.5

OPERATING AND GENERAL STATISTICS

Year	Generator Ratings Nameplate Kw Thousands	Generator Ratings Capability Kw Thousands	Peak Customer Demand Kw Thousands	Net Output Kwhr Millions	Annual Load Factor %	# Coal Per Kwhr Net Output	Average Fuel Cost ¢ Per Million Btu	Btu Per Kwhr Net Output	Operating Employes Average Number for Year	Operating Employes Average Wage $ Per Hour
1945	1,234	1,336	1,034	5,163	57.0	1.01	21.6	12,660	6,390	1.29
1946	1,234	1,336	1,192	5,291	50.7	1.02	23.7	12,715	7,385	1.50
1947	1,309	1,420	1,289	6,368	56.4	1.03	27.9	12,810	8,096	1.60
1948	1,309	1,420	1,351	6,922	58.3	1.04	32.7	12,880	8,350	1.69
1949	1,409	1,545	1,412	7,054	57.0	1.01	31.7	12,900	8,534	1.78
1950	1,508	1,669	1,603	8,188	58.3	.98	30.9	12,450	8,748	1.84
1951	1,708	1,895	1,643	8,982	62.4	.97	30.3	12,360	8,904	2.04
1952	1,708	1,895	1,867	9,351	57.0	.95	30.3	12,190	9,073	2.20
1953	1,958	2,195	1,945	10,709	62.9	.94	30.5	12,100	8,791	2.31
1954	2,208	2,535	2,145	10,625	56.6	.89	29.0	11,340	8,750	2.41
1955	2,333	2,550	2,367	12,726	61.4	.88	28.0	11,080	8,497	2.51
1956	2,593	2,841	2,410	13,069	61.7	.86	30.2	10,770	8,347	2.62
1957	2,853	3,101	2,470	13,792	63.7	.86	31.6	10,660	8,217	2.76
1958	3,174	3,426	2,558	13,010	58.1	.83	31.7	10,260	7,887	2.93
1959	3,511	3,773	2,689	14,493	61.5	.81	31.3	9,990	7,654	3.09
1960	3,511	3,762	2,741	15,445	64.1	.81	30.4	10,060	7,475	3.24
1961	3,836	4,087	2,884	15,686	62.1	.79	30.3	9,740	7,406	3.37
1962	3,829	4,099	3,079	17,120	63.5	.79	29.1	9,720	7,230	3.49
1963	3,829	4,099	3,281	18,632	64.8	.80	28.6	9,800	7,159	3.60
1964	3,866	4,111	3,616	20,375	64.2	.80	28.1	9,920	7,147	3.73
1965	3,923	4,316	3,975	22,375	64.3	.82	27.7	10,130	7,138	3.82
1966	4,259	4,619	4,273	24,704	66.0	.86	28.4	10,490	7,532	3.95
1967	4,345	4,903	4,422	25,611	66.1	.87	29.1	10,660	7,701	4.04
1968	5,039	5,404	4,835	28,039	66.0	.86	29.6	10,460	7,690	4.39
1969	5,616	5,733	5,201	30,855	67.7	.86	31.5	10,360	7,682	4.67

OPERATING AND GENERAL STATISTICS

Year	Service Area Square Miles	Population Served (Estimated) Thousands	Annual Revenue per Domestic Customer	Annual Kwhr Use per Domestic Customer	Major Domestic Appliances in Service at Year End (Thousands of Customers)			
					Ranges	Water Heaters	Refrigerators	Clothes Dryers
1945	7,587	2,942	42.35	1,399	150	12	500	
1946	7,587	2,945	43.46	1,513				
1947	7,587	3,069	45.93	1,641				
1948	7,587	3,102	49.01	1,746				
1949	7,587	3,184	57.03	1,845				
1950	7,587	3,414	61.42	1,992	225	52	780	26
1951	7,587	3,552	65.38	2,168				
1952	7,587	3,662	68.72	2,317				
1953	7,587	3,772	75.51	2,489				
1954	7,587	3,828	76.11	2,656				
1955	7,587	3,950	78.40	2,776	350	104	1,000	150
1956	7,587	4,120	83.45	2,986				
1957	7,587	4,224	85.85	3,120				
1958	7,587	4,237	87.58	3,211				
1959	7,587	4,228	92.41	3,412				
1960	7,587	4,243	94.46	3,514	410	148	1,150	175
1961	7,587	4,294	97.36	3,653				
1962	7,587	4,332	99.34	3,753				
1963	7,587	4,387	102.28	3,889				
1964	7,587	4,456	106.85	4,102				
1965	7,587	4,572	109.71	4,335	550	174	1,265	310
1966	7,587	4,695	114.95	4,622				
1967	7,587	4,786	119.47	4,882				
1968	7,587	4,844	126.44	5,296				
1969	7,587	4,858	133.44	5,697	625	198	1,389	340

Appendix C

OFFICERS 1951–1970

	Tenure of Office	
	FROM	TO
Chairman of the Board		
Walker L. Cisler	5-4-64	*
Presidents		
Walker L. Cisler	12-1-51	5-4-64
Donald F. Kigar	5-4-64	7-1-67
Edwin O. George	7-1-67	4-20-70
William G. Meese	4-20-70	*
Executive Vice Presidents		
Charles R. Landrigan	7-1-54	9-1-61
Donald F. Kigar	9-1-61	5-4-64
Howard R. Stevenson	4-21-69	8-1-69
Harvey A. Wagner	4-21-69	2-1-70
Executive Vice Presidents for Marketing		
Edwin O. George	4-17-67	7-1-67
Howard R. Stevenson	7-1-67	4-21-69
George L. Lahodny	4-21-69	*

*Still in office as of January 12, 1971

Executive Vice Presidents for Production

George A. Porter	4-17-67	9-1-67
Harvey A. Wagner	9-1-67	4-21-69
William G. Meese	4-21-69	4-20-70
Harold C. Reasoner	4-20-70	*

Executive Vice President for Finance

Robert W. Hartwell	1-1-70	*

Administrative Vice President

Chester F. Ogden	5-4-64	*

Senior Vice Presidents

George A. Porter	4-19-65	4-17-67
Eldred H. Scott	4-19-65	5-1-68
Edwin O. George	4-19-65	4-17-67

Vice Presidents

George A. Porter	12-1-52	4-19-65
Charles R. Landrigan	1-1-53	7-1-54
Thomas E. Hurns	5-2-55	5-1-58
Eldred H. Scott	5-2-55	4-19-65
Edwin O. George	9-1-56	4-19-65
Arthur S. Griswold	12-1-57	5-1-68
Donald F. Kigar	12-1-57	9-1-61
Sylvester F. Leahy	12-1-57	5-1-68
Chester F. Ogden	12-1-57	5-4-64
Marshall Pease	5-4-64	6-1-69

*Still in office as of January 12, 1971

Howard R. Stevenson	4-19-65	7-1-67
Harvey A. Wagner	4-19-65	9-1-67
William G. Meese	9-1-67	4-21-69
George L. Lahodny	7-1-67	4-21-69
James P. McCormick	4-15-68	*
Robert E. Schwab	5-1-68	*
Robert W. Hartwell	1-1-69	1-1-70
Robert W. Lundgren	6-1-69	*
George J. Brown	1-1-70	*
John W. Drummond	1-1-70	*
Harold C. Reasoner	1-1-70	4-20-70
Paul A. Duker	1-1-70	*

Secretaries

Thomas E. Hurns	12-1-51	5-1-58
Anthony G. Maihofer	5-1-58	7-1-63
Ray J. Plourde	7-1-63	11-1-68
Harris R. Symes	11-1-68	*

Treasurers

Edward M. Spencer	3-1-57	5-1-69
Phillip D. French	5-1-69	*

Controllers

Eldred H. Scott	4-14-53	3-1-68
Robert W. Hartwell	3-1-68	*

General Accountants

Everett E. Roll	5-7-56	12-31-68
Robert O. Wagner	1-1-69	*

*Still in office as of January 12, 1971

Auditors

Charles J. Ghesquiere	4-1-46	12-1-62
Harris R. Symes	12-1-62	11-1-68
Arnold J. Benes	11-1-68	*

DIRECTORS

	Tenure of Office	
	FROM	TO
James F. Fogarty	(2-2-20	9-20-20
	(1-15-34	8-25-60
Alfred C. Marshall	9-20-20	2-9-45
E. W. Rice, Jr.	9-20-20	11-25-35
Harrison Williams	11-15-20	11-10-22
F. N. B. Close	10-20-22	9-10-23
Edwin Gruhl	11-10-22	1-22-33
B. A. Tompkins	(9-10-23	6-17-41
	(9-15-42	*
James S. Holden	1-18-24	2-14-44
Standish Backus	12-17-25	2-5-34
Royall Victor	11-19-25	5-30-26
John Foster Dulles	9-10-26	3-19-40
Frank L. Dame	2-16-33	12-30-33
Wesson Seyburn	2-5-34	2-14-44
Edward T. Gushée	7-20-36	5-16-39
James W. Parker	9-17-40	12-30-57
George L. Burr	2-3-41	11-19-56
Silas A. Strickland	2-3-41	6-17-43
Charles T. Fisher, Sr.	5-11-43	10-21-47
Reuben B. Crispell	6-17-43	11-21-55
Frank L. Conrad	2-14-44	6-15-48
Harry B. Munsell	2-14-44	1-15-46
Prentiss M. Brown	5-16-44	*
Russell W. Symes	2-20-45	3-19-56

*Still in office as of January 12, 1970

Ralph T. McElvenny	1-15-46	9-20-49
Charles T. Fisher, Jr.	10-21-47	4-14-58
George J. Burke	9-20-49	10-3-50
George R. Fink	3-15-49	11-21-55
Robert R. McMath	3-15-49	1-2-62
Walker L. Cisler	9-18-51	*
Daniel F. Gerber	11-21-55	4-18-66
William C. Foster	2-20-56	10-16-61
Raymond T. Perring	3-19-56	*
H. Glenn Bixby	7-16-56	*
Harlan Hatcher	7-16-56	*
Richard C. Andreae	3-18-57	*
George M. Holley, Jr.	2-17-58	*
Lawrence P. Fisher	6-23-58	9-3-61
Frederick H. Mueller	2-20-61	*
Charles T. Fisher III	9-18-61	*
Robert F. Bacher	11-20-61	*
Joseph L. Hudson, Jr.	2-19-62	*
Kenneth D. Nichols	2-19-62	*
Donald F. Kigar	5-4-64	7-1-67
James Boyd	4-18-66	*
Edwin O. George	7-1-67	*
Alan E. Schwartz	4-21-69	*
William G. Meese	4-20-70	*

*Still in office as of January 12, 1971

Appendix D

MEMBER COMPANIES

Atomic Power Development Associates, Inc.

1955–1965

Alabama Power Company
Allis-Chalmers Manufacturing Company
Atlantic City Electric Company
Babcock & Wilcox Company, The
Baltimore Gas and Electric Company
Bendix Corporation, The
Boston Edison Company
Central Hudson Gas & Electric Corporation
Cincinnati Gas & Electric Company, The
Cleveland Electric Illuminating Company, The
Commonwealth Associates, Inc.
Connecticut Light & Power Company, The
Consolidated Edison Company of New York, Inc.
Consolidation Coal Company, Inc.
Consumers Power Company
Delaware Power & Light Company
Detroit Edison Company, The
Ford Motor Company
General Motors Corporation
General Public Utilities Corporation
Georgia Power Company
Gulf Power Company
Hartford Electric Light Company, The
Indianapolis Power & Light Company
Jackson & Moreland, Inc.
Jersey Central Power & Light Company
Leeds & Northrup Company
Long Island Lighting Company
Metropolitan Edison Company
Mississippi Power Company
NEGEA Service Corporation
New England Power Company

New Jersey Power & Light Company
New York State Electric & Gas Corporation
Niagara Mohawk Power Corporation
Pennsylvania Electric Company
Philadelphia Electric Company
Potomac Electric Power Company
Public Service Electric & Gas Company
Rochester Gas and Electric Corporation
Southern Services, Inc.
Toledo Edison Company, The
United Engineers & Constructors Inc.
Vitro Corporation of America
Wisconsin Electric Power Company
Wisconsin Power and Light Company

MEMBER COMPANIES
Power Reactor Development Company
1968

Allis-Chalmers Manufacturing Company
Babcock & Wilcox Company, The
Burroughs Corporation
Central Hudson Gas and Electric Corporation
Central Research Institute of Electric Power Industry (Japan)
Cincinnati Gas and Electric Company, The
Columbus and Southern Ohio Electric Company
Combustion Engineering, Inc.
Consumers Power Company
Delmarva Power and Light Company
Detroit Edison Company, The
Holley Carburetor Company
Iowa-Illinois Gas and Electric Company
Long Island Lighting Company
Philadelphia Electric Company
Potomac Electric Power Company
Rochester Gas and Electric Corporation
Southern Services, Inc. as representative of
Southern Company, The

Alabama Power Company
Georgia Power Company
Gulf Power Company
Mississippi Power Company
Toledo Edison Company, The
Westinghouse Electric Corporation
Wisconsin Electric Power Company

Acknowledgements and Sources

In connection with the notes which follow, I have discussed the documentary sources on which this study rests. Let me, with deep appreciation, acknowledge another and unusual source of information. For two years I sat as a member of Management Council, listening to the discussions of problems and policies, and to the announcement of decisions. I have been received with uniform courtesy, and dealt with with obvious candor.

It is not often that a historian has the opportunity to discuss with his subject the basis of his decisions and the nature of the alternatives available. Any historian can point out the threat to objectivity which is present in this kind of involvement. I would be less than honest if I did not acknowledge that these men have won my admiration for the skill with which they get the job done, and my thanks for the friendship which many of them generously offered. Let me be equally clear on another point; the members of the Company to whom I talked understood and respected the obligations which scholarship imposes upon me.

Some of them found records which were in personal possession, and made them available. Interviews and questions must have interrupted many a busy schedule. In some cases, I asked for a critical reading of a draft of the manuscript covering events or a technology on which they had special competence. It is a pleasure to acknowledge my personal obligation.

Among the officers my especial thanks is due
To Robert W. Hartwell, Chester F. Ogden and George L. Lahodney,
To Harold C. Reasoner, John W. Drummond and George J. Brown,
To Harris R. Symes, Phillip D. French and James P. McCormick,
To a number of officers now retired; Eldred H. Scott, Edward M. Spencer, Howard R. Stevenson, Marshall Pease, Harvey A. Wagner and Ray J. Plourde.

In a number of cases, officers made available to me staff personnel to suppliment their personal assistance. My thanks are thus due

To Robert W. Lundgren and John F. Anderson,
To Robert E. Schwab and Alfred L. Kasameyer,
To Paul A. Duker, William R. Milby and Edward J. Hurley,

To Walter J. McCarthy Jr. and J. Douglas Elliot.

The professional competence in the staff of The Detroit Edison Company is at a high level. To the following persons, I have expressed my thanks for the assistance and my appreciation of the skills and knowledge involved.

To Lois Brandenburg, Ruth R. Rush and Grace M. Pacak,
To Frederick W. Braga, Robert Paoletti and Kenneth Locke,
To John U. Harkness, Walter Doyle and Margaret L. Furlong,
To Paul Van der Meer, Wilber J. Pierce and Arthur E. Bush.

All charts and maps are by Company artists, and most of the photographs as well. All are in Company possession. The picture of Sarah Sheridan is from a portrait by John Coppin. For this material, my special thanks go to William B. Milne, Edward T. Smith and Dale A. Collins, and the staff.

Notes

A DESCRIPTION OF THE SOURCES

PART ONE

This work rests on the voluminous official records of the Company, and on the related correspondence, reports, minutes and notes. Most of this material is in the Records Center of the Company, or in the appropriate office. Since the documentation presented here does not show the extent or nature of these materials, a brief explanation is in order.

In anticipation of the celebration of its fiftieth birthday, I was invited by the late Charles Landrigan, then controller, to do a biography of The Detroit Edison Company. Neither of us was aware of the dimensions of the proposed undertaking. The work was published with the title *Kilowatts at Work: A History of The Detroit Edison Company* (Wayne State University Press), 1957.

With the exception of the final chapter, that volume carried a full and detailed documentation. Substantially, then, this earlier work, with full documentation and with complete access to Company records closes in 1946, with a single final chapter, undocumented, which described generally certain on-going activities.

When I accepted the invitation by Walker Cisler, Chairman of the Board of Detroit Edison, to undertake this second commission, the question of the relationship between these two works became important. Since the older book is out of print, there was no point in designating this as Volume II of a set. It is obvious that since this book is to stand on its own feet, it must contain enough background to make the last quarter century understandable. The decision was made, accordingly, to use Part One to provide that relevant material.

The documentation of such an abbreviated narrative is especially difficult. Let me refer the reader directly to the appropriate sections of the older work, where some 28 pages of notes and references support the narrative into the first half decade after 1946.

In the absence of conventional documentation, it is appropriate at least to describe the available sources. The Detroit Edison Company was organized in 1903. The new company was direct heir to The Edison Illuminating Company of Detroit, which had existed since 1886; records of that company survive in the archives of the successor organization. Because

The Illuminating Company—by purchase or otherwise—had acquired several earlier electric companies, large fragments of their materials also have survived. These competitive companies differed in the patents they proposed to use, and the technology played a large part in the success or failure of the enterprises.

After 1903, the records of the Company are continuous and full. The Washington Boulevard Building, which was the first office headquarters, still survives. Records were moved to the archives when the present General Office Building was occupied. There has been no accidental destruction of records by fire or flood. It is possible that some material could have been removed, but there are no obvious gaps which would indicate this.

For these earlier years, the records from the office of the president are most important. From 1896 to 1912, Alex Dow was manager and vice president, and from 1912 to 1942 president; in his day, the significant decisions were made by him or rested on material which reached his desk in detail. The files of the president contain letters, carbons of replies, telegrams, memoranda of meetings or telephone conversations, recommendations and reports from associates and other documents; at the time, I estimated that I had read a third of a million documents in these files. These records were not then but are now a part of Records Center.

In general, for the period covered in Part One, the other materials seem to be relatively complete. In Records Center are most but not all of the business records, contracts, agreements, often with notes or explanatory correspondence. These include property records, tax records, reports of the public accountant and even reports on stock ownership and transfer. The Minutes of the shareholders meetings, and of the board of directors are in the office of the secretary. Officers of the Company, including the secretary, the treasurer and the controller, made available to me materials in their respective offices which had not been transferred to Records Center.

Special reference should be made to the controllers office, and to the statistical division which is in that office. When I was working on the earlier history of the Company, the office library and file contained a great deal of otherwise unobtainable material. The elaborate statistical material in the appendix of this volume is provided by the statistical division.

CHAPTER ONE *Beginnings in the Industry*

1. The rivalry between the National companies with differing patented technologies can best be followed in Harold G. Passer, *The*

Electrical Manufacturers, 1875–1900 (Cambridge: Harvard Univerisity Press, 1953). A work dealing specifically with the local situation in Detroit in the same early years is William R. Lane, *A History of Electrical Service in Detroit.* (Detroit Bureau of Governmental Research, 1937).

2. The account of the developing technologies is of interest; of even greater significance is the emergence of a conception of responsibility as a public utility, as contrasted with the familiar behavior of the competitive vendor. *Kilowatts at Work* discusses both points in greater detail.
3. In *The Electrical World,* the most significant trade journal of the industry, the issue of September 10, 1887 presented an elaborately illustrated description of the new Washington Boulevard Building, which later (October 29) it declared to be second only to a New York power plant in size.
4. The turbine, to achieve full efficiency, required the flow of cooling water to the condenser unit. The available water front selected was in the village of Delray, immediately down river from Detroit. Almost at once, Delray was annexed to the city.
5. The literature on Ford and the Ford Motor Company is voluminous. The three volume work by Allan Nevins is the standard, and is probably the most successful in interpreting what came to be called the Detroit Method.

Chapter Two *The Appearance of Crisis*

The Company requested a ruling supporting its contention that it was not a held company, within the meaning of the Holding Company Act. The resulting investigation was by the staff of the Securities and Exchange Commission, in 1939. With substantially unlimited funds, its people combed the records of the banks and other companies with which Edison did business, and presented in evidence documents and correspondence which would not normally be available to historical research. The entire proceedings, including testimony and exhibits were in the possession of the controller, and have recently been placed in Records Center.

This manuscript collection, now assembled in six large volumes, supplements the Company's own records. The issue was essentially a historical one, hanging on the question of fact in operation. The relevance to any discussion of Company tradition and customary procedure is clear.

1. The traditional statement of dividends in percentage terms was common when, as in the case of The Detroit Edison Company, par for the stock was one hundred dollars. When, on March 21, 1941, the par was reduced to twenty dollars by a five for one stock split, the payment of $1.20 as a reestablished dividend was the same six percent of par.
2. Reference here is made to the proceedings and the documentary evidence already described.

Chapter Three *The Board and its Chairmen*

1. The names and the effective dates of the members of the board to 1970 are in the appendix to this volume.
2. Prentiss Brown, in his own summary of his career in the Company, listed first this solution of the problem of finance. The quotation is from this summation, as printed in *Synchroscope* (July, 1954). An earlier issue of this employe magazine, (June, 1954) had carried an extensive biography of Brown on the occasion of his retirement.
3. To be accurate, one exception should be noted. During the early part of World War II, the dividend payment was increased by five cents per share for three consecutive quarters, and was then returned to the earlier amount.
4. Before the second meeting, Eldred Scott reported that 68,076 of the shareholders were also customers; most of these holdings were small. At that time, 2,140 persons were at once shareholders, customers and employes. *Synchroscope,* July, 1953.
5. The statement appears in the review of his career, a message to the Company on his retirement. *Synchroscope,* July, 1954.
6. The returns to the Company had been distorted in the four critical prewar years by the sharp recession in the automobile industry. For the years 1936–39 inclusive, the figures were 3,079,000, 3,920,000, 2,019,000, and 2,888,000 cars produced.
7. The receipt of the official permission to export was reported to the board November 10, 1953.
8. The appointment of Brown as special counsel was approved by the board. *Board Minutes,* May 6, 1957.
9. The formal ceremonies at the bridge were scheduled for June 26–28. The board meeting was called for June 23. *Minutes of the Board,* April 21, 1954.

CHAPTER FOUR *The Company and its Presidents*

1. A long time friend and associate, Charles M. Drake, wrote a detailed biographical sketch of Parker on his retirement. *Synchroscope,* December, 1951.
2. W. B. Hurley, from Sales, worked with Parker on the Committee for Economic Development. More than seven thousand responses were received from the eight thousand questionnaires sent out to the three county area. The details of the method as well as the results are given in the annual report of the sales department for 1945.
3. The machine had been ordered, but delivery was held up by the War Production Board. The chief engineer in the Company assured the board that the demands of the war period could be met, but that the machine would be needed in the postwar demand. The machine would be installed at Marysville as soon as the board released it. S. M. Dean to M. L. Perlette, March 12, 1945.
4. The announcement was reprinted in *Synchroscope* January, 1950. The peak had occurred on December 15, but there had been no crisis. Parker referred to the announcement as a "report on our stewardship."
5. The Detroit Edison Company, *Annual Report* for 1952. The union represented about 3,700 members. The settlement of the longer strike is reported in the *Annual Report* for 1955.
6. The figures given are from the Question Box in *Synchroscope,* August, 1950.
7. The Detroit Edison Company, *Annual Report,* 1956. The first conception of the pension was as a supplement to Social Security payments. In 1956, however, when both costs and benefits were increased, the Company decided not to modify the Company appropriations. *Minutes of the Board,* April 17, 1956.
8. The Detroit Edison Company, *Annual Report,* 1951.
9. Address of Cisler to the Power Club, as printed in *Synchroscope,* March, 1955. This was a serious discussion of the problem of promotion and job transfer as the theme of the featured program. To emphasize the actual mobility, Cisler pointed out that in the next five years, 714 employes would reach retirement age.
10. Sylvester Leahy, *Synchroscope,* April, 1954.

PART TWO

In the period covered by Part Two, materials in manuscript form are less available. There remains in the Company nothing comparable to the file of the president in the days of Alex Dow. This reflects the change in the office of chief executive, and in some degree in the Company itself. Official records of the corporation, retained by the several officers, continue to provide basic minutes on actions taken, but are seldom more illuminating. The correspondence, memoranda and reports in Records Center tend to become fragmentary. In part this is a result of size and the multiplication of officers. The deposit of the noncurrent material was not formally prescribed, and it was sometimes discarded, especially when files were cleared on retirement. Members of management took too seriously some efficiency expert who calculated the cost of filing a sheet of paper.

One collection of manuscripts proved to be of special importance. In 1954, when he succeeded Brown as executive officer of the Company, Cisler began a series of special reports to the members of the board of directors. It is a little ironic that even the first several volumes of this series have been destroyed. This *Letter to the Board* is a monthly communication distributed to the members usually in anticipation of a meeting, though on occasion a special letter of information was transmitted. Rigorously contemporary, the letter presented statistical material, discussed problems, considered alternatives, and on occasion argued for the position management had taken on a given issue. The *Letter to the Board* is an especially valuable aid in understanding formal board actions, which normally were unexplained in the *Minutes*.

The sales department, currently the marketing department, has for historical reasons had a rather autonomous character, with widely spread activity. The annual report of the department is not printed, and the typescript copy varies in form and content from year to year but it is usually comprehensive. The discussion of actions is associated with extensive presentation of statistical information. In the annual report, and in the other reports from this department on special topics, can be traced the relationship of the Company with the several industries of the metropolis, and with the general customer. Some other departments issue reports, summaries or statistical material in tabular form, and when these have been used they will be noted either in the text or in the notes.

Fortunately, the Company tends to emphasize committee consideration, and that means a number of copies of minutes and significant documents. With some searching, I was usually able to find some office in which committee records had escaped destruction so far. Especially after the administrative tightening which is described in Chapter VII, the committee minutes and correspondence of System Planning, Capital Expenditure, Load Forecast, and of certain special committees became vital.

In the rather simple and routine development of plants and system, the files of the *Synchroscope* were of great use. This employe magazine became in this period less a scientific and educational journal, and more a journalistic record of Company events, presented with pictures and often with some technical detail. Appointments and retirements were noted with biographical material not easily available elsewhere.

There were several important employe organizations; the officers of the Company were likely to announce or defend policy before these significant groups, or to talk about Company plans. One very useful feature of the *Synchroscope* was the publication of speeches and talks given inside the Company. One important series is a report of the local Detroit annual shareholders meeting, which often went beyond the scope of the *Annual Report* in the discussion of policy.

The *Synchroscope* developed a feature which, entitled simply "The Question Box," became important. Communications from employes, even unsigned and sometimes improper ones, received careful and candid answers from responsible Company officials. The feature was first used early in the administration of Parker, when J. H. Walker was made vice president with special responsibilities. After his untimely death, custody of the Question Box went first to Walker Cisler, then Charles Landrigan and Chester F Ogden, as each in turn became executive vice president. The questions reveal the concerns of the employe, and the answer frankly discussed the policy or procedure which was involved.

CHAPTER FIVE *Opportunity and the New Leadership*

1. The impact of the regulatory acts was described in Chapter Two. The application of the several acts was by intention disruptive of the traditional and established relationships.
2. The *Letter to the Board* earlier than this date may not have been formalized. The file in the office of the secretary notes the existence of earlier series of letters, but the file available is from this date; thereafter, the series is continuous.
3. An extensive biographical note was published in the employe magazine announcing the promotions. They include personal data, the Company career and the record of professional recognition. These appointments were noted in the *Annual Report,* 1952.
4. In its earlier history, the Company had no organization charts, in

large degree because Alex Dow preferred not to be restricted in assigning responsibilities. In 1942, the older organizational lists were for the first time given a kind of schematic treatment which was shown on a chart. The first Table of Organization is dated May 20, 1952, and shows both rank and reporting responsibilities. Records Center has a complete file of these lists and charts.

5. In what was referred to as his presidential address to the Power Club, Cisler indicated the results he hoped to obtain. "The managers. . . . are delegated full responsibility for supervising all of the functions within their jurisdictions and for making the decisions for the conduct of their departments. In this way we have established managerial authority as close as possible to the actual work. . . . Our management structure is clear and direct." *Synchroscope* March, 1955.
6. *Synchroscope,* July, 1952. The changes in reporting responsibility and in organization were announced later in General Orders, effective November 1, 1953.
7. One significant index of competence is recognition by appropriate professional organizations. In 1956, the names of 42 Detroit Edison employes and officers appeared in the list of officers and committee members of the Edison Electric Institute (EEI). Walker Cisler was at that time a member of the board of directors and of the Atomic Power Committee.
8. The report of the Wagner committee was given wide publicity within the Company. It was printed in *Supervisor's News* August 18, 1960. The *Letter to the Board* August 15, 1960, also stressed the importance of the recruitment and development program.
9. The problem of recruitment continued to be of concern to management. Applicants for full employment as engineers dropped from 200 in 1961 to 139 the following year. Declining registration in engineering colleges was one probable explanation, and Cisler announced his intention to attempt additional methods of recruitment. *Letter to the Board,* April 10, 1962.

Chapter Six *The Problem of Power Production*

1. D. J. DeBoer to Samuel M. Dean, September 1, 1942 ff. The correspondence between the two men continued; as the reports became more formal they carried the title *"Load and Capability for Area II."* The shift from the report on current situations to consideration of future contingencies came slowly.
2. S. M. Dean to DeBoer October, 1943. Dean described his experi-

ence in attempting to coordinate the activities of these plants, and the narrow margin on which wartime Detroit was forced to operate.

3. Cisler's report indicated plans which included adding Trenton Channel #7 and #8, and Conners Creek #15 and #16. Including overload capability, the Company expected to move from 1,420 mw to 1,872 mw in the years 1948 to 1952 inclusive. The EEI committee was the *Special Committee on Load Capability.*
4. On June 12, 1947, at the request of Cisler, H. S. Bennion, director, by letter thanked the members of the industry for their cooperation, and reported a 95 percent participation.
5. For a number of years, A. Douglas Campbell and Arthur S. Griswold either singly or together compiled the semiannual report for the committee, which won wide acceptance. For example, in 1951, it included figures from fifty representative power systems, power supply areas and manufacturing establishments, a participation by 98 percent of the American industry. The report for January, 1951 indicated that 368 steam and 130 hydro turbines were planned as additions before the end of 1953, the four years covered by the report.
6. *Synchroscope,* March, 1949.
7. This statement is subject to qualifications. The "extensions" at the older plants were in fact totally new, and the presence of the older plants on the same grounds was only to a small degree a limitation. Delray III was a complete and independent plant.
8. The St. Clair plant was the first totally new plant within the current generation. The excitement its construction created throughout the Company is indicated by the detailed reporting of its progress in the *Synchroscope,* 1949 to 1955.
9. By special letter, Cisler invited the members of the board to be present February 20, 1956, when board member Fisher would turn on "the largest generator in the world." On October 20, 1958, the board met again at the River Rouge plant to turn on #3.
10. *Synchroscope,* November, 1956.

Chapter Seven *Riches to Rags and Back*

1. The emphasis in this chapter is on the Company's reaction to the swings of the general economy. In this study, the *Letter to the Board* was of singular importance. As a monthly commentary on the actions taken to meet the changing situation, the material is thus quite contemporary, very frank, and exceedingly detailed.

2. The recommendation of the committee was transmitted in the *Letter to the Board,* January 11, 1961.
3. The *Annual Report* for 1959 noted that less had been expended in capital funds than in any of the previous eleven years. Specifically, the *Report* for 1960 pointed out that construction costs were at their lowest in the Company's recent history, and the conviction was expressed that funds internally generated would be adequate for construction for several years to come.
4. This address was presented to the American Society of Mechanical Engineers, Maintenance and Plant Engineering Conference, St. Louis, April 25, 1960.
5. The number of employes is normally reported in the *Annual Report,* usually with a comment on trends. In the *Letter to the Board,* January 11, 1961, Cisler reviewed the experience of the decade; the figures given are from this statement.
6. Almost a year later, in the *Letter to the Board* October 11, 1961, the president reported a further reduction in nine months of 232 employes.
7. The System Development Committee and the Capital Expenditures Committee promptly moved to positions of recognized importance. They reviewed the budget for 1962, and their statement of approval was added to that of the vice president and controller before it went to the board for action. In 1963, the approval of the Administrative Systems Advisory Committee was also indicated. The earlier practice of consultation was being formalized in these actions.
8. Albright to Cisler, October 14, 1953. The immediate problem of the Company was with ordering, billing and collecting of customer accounts, estimated at the time at one hundred million manual or machine operations per year. The report of the task force (October 11, 1954) proposed reorganization which would allow the extensive use of machines. Stevenson and Elliott to the management committee, C. R. Landrigan, F. A. Compton and T. E. Hurns, December 16, 1954.
9. The work of the committee within the Company was supplemented by three elaborate reports from the New York consultant firm of Cresap, McCormick and Paget. These reports evaluated the proposal for the use of electronic equipment on customer accounts, and the recommendation for an internal organization which would best extend the use of such equipment.
10. Charles R. Landrigan to Cisler, November 14, 1960. The letter, with the approval of the president, was included in the *Letter to the Board.*

CHAPTER EIGHT *Lines and Systems*

Reference is made to a small volume *The History of the Overhead Lines Department,* by Clifford P. Wells (privately printed, The Detroit Edison Company, 1953), which contains some supplemental chapters on specific activities by other competent persons. The account includes changes in organization, the development of specialized functions and the changing personnel of the department.

Another employe magazine, *Supervisor's News,* appealed to a more restricted audience than did the *Synchroscope;* it tended to be more technical in content, but was also designed to interest and inform.

1. No effort has been made to reconcile minor inconsistencies in voltage ratings as reported. In general, improved technology and equipment allowed the voltages to move up with more exacting performance standards.
2. The Detroit Edison Company, *Annual Report,* 1951, noted the completion of the 120,000 volt line between Delray and Conners Creek stations.
3. The figures are from an article in *Supervisor's News* entitled "Kilowatts to Customers," July 12, 1956, in which the whole problem of higher voltages was discussed.
4. *Supervisor's News,* November 16, 1956. A more formal announcement appeared in The Detroit Edison Company *Annual Report,* 1954.
5. Cisler was much impressed by Russian technology, and especially by what he called their "massive approach" to their problems. *Synchroscope* October, 1961.
6. The *Report* of the department committee to the System Planning Committee, February 23, 1965.
7. The award, presented April 9, 1964, was noted in *Letter to the Board,* March 10, 1964; such notice was unusual. The establishment of the service in its developed form was mentioned in The Detroit Edison Company *Annual Report,* 1962.
8. The windstorm was discussed in detail in *Letter to the Board,* September 10, 1968. For the storm of December 7, see *Synchroscope,* January 14, 1969.

CHAPTER NINE *Sales and the Domestic Revolution*

The concentration in the sales department of full control of customer services and related Company functions has given it a position probably unique in the industry.

The annual report of that department was thus both an explanation and a defense. To management, and to the rest of the Company, the report stated problems, programs and accomplishments, and defended actions taken or recommended. The value of the report to the historian is obvious; citation of the individual documents will indicate the richness of the whole collection.

1. *Supervisor's News* August 21, 1959.
2. Compton reported in 1946 that 40 percent of the new business in the commercial division was the result of direct solicitation for business.
3. Edwin O. George was assistant commercial manager when he was invited to address the Seventeenth Annual Sales Convention of the Edison Electric Institute meeting in Chicago, April 2–5, 1951. In this address he stressed the traditional concern with the public, and insisted on industry responsibility not only for its own service, but for that also of the appliances it sold or recommended. The idea that any sale was a good one was one which George and his associates were in due time to challenge.
4. Market research was the study of what the customer might be persuaded to buy. Load research was the study of the use by customers, with reference to the time of normal use. Howard Stevenson was invited, on several occasions, to present the results of the Company study to national professional organizations. On September 8, 1958, he spoke to the Ebasco Workshop in New York on "Selling to Improve Load Factor." On February 10, 1959, he addressed the American Marketing Association, Fifth Annual Public Utilities Seminar, meeting in Detroit, on the subject "Load Research as an aid to Market Planning." In this paper, Stevenson urged the use of such load research in the redesigning of rate structures.
5. In the *Letter to the Board,* September 11, 1967, Cisler reported that the ICC license had been delayed, but was issued August 31. By September, 1968, 100,000 of these devices had been installed and the Company was pleased with the operation. The ability to "shave the peak" by temporary cut off could save as much as $400,000 in fuel alone.
6. This discussion of the electric water heater deliberately avoids the highly complex issue of rates. Between 1955–69, the Company, with the approval of the Commission, had tried a sequence of adjustments. Understandably, customers who had accepted the invitation to purchase felt that they had a commitment, and the new conditions or rates often applied only to new customers. The result was a series of rates or systems of rate adjustments

which operated at the same time, and a confusion which in 1969 led to a challenge of the radio control system before the Commission.

7. The summer peak, when it at length arrived in Detroit Edison territory, changed the situation completely, and justified the consideration of a special promotional rate for off peak heating load.
8. Edwin O. George emphasized the new situation in his address "Electric Heating Has Come of Age." He spoke on behalf of the industry to the St. Louis meeting (December 3, 1959) of the National Warm Air Heating and Air Conditioning Association.
9. The Detroit Edison Company, *Annual Report,* 1948, stated that 98 percent of the farms of The Detroit Edison territory had electricity, and the balance could have it on application.
10. *Synchroscope* April, 1960.

CHAPTER TEN *In the Service of Industry*

The annual reports of the sales department continue to be a most important source of information; this is especially true in matters concerning industrial use since they were not given public notice or announced in the press. Several statistical series exist covering both revenue and demand in the service to industry.

1. The U. S. Census of Manufacturing, 1947, as reported in the *Detroiter,* October 16, 1950.
2. In strict accuracy, this should be called the primary sales. The reason for the inclusion of some large commercial accounts in this classification will be discussed later; with this exception, the primary sales were to industrial customers.
3. The load factor is the ratio between peak demand and average utilization, expressed in percentage terms.
4. At Cisler's request, Hurley accepted appointment on the governor's commission relative to defense contracts. The sales department report for 1952 attributed to its efforts the success of Michigan in winning third place among the states in total contracts. The total was $7.5 billion, for 11.2 percent of the nation's total. At the time there was a surplus of jobs in the state.
5. In 1953, the Area Development division was established within the sales department, with Vincent Madison as division manager.
6. The record of the developing steel industry in the Detroit area is

a separate story. These plants were relatively new, and their rapid expansion in production facilities allowed them to adopt the most modern methods. In this research and development, men from Detroit Edison sales were constant participants.

7. In 1948, there were 1,735 customers in the downtown area; an increase in that year of fifty was the largest increase in years. The heating service was connected to 95 of the downtown buildings. The Detroit Edison Company, *Annual Report* for 1948.
8. *Synchroscope,* November, 1957
9. *Synchroscope,* November, 1959.
10. *Synchroscope,* April, 1963. The figures include the production of the Detroit Public Lighting Commission.
11. The Company's *Annual Report,* 1958.
12. The proposal from Edwin O. George, then vice president for Sales, to Cisler, was transmitted in the *Letter to the Board,* January 15, 1963. The contract, as negotiated, was approved by Pennsalt on February 6, and by Edison on March 18. The actual takeover was then scheduled for April 1, 1964.
13. Cisler described the negotiations in the *Letter to the Board* June 1, 1964. It was known that a good deal of capital would be required to modernize these old plants. Wyandotte's directors approved the sale on October 27, to become effective early the next year. The assumption of operating responsibility was postponed by a strike in the chemical plant.
14. The Ford Motor Company was on the list, but the remainder of the companies were small, with limited demand for electrical energy. *Letter to the Board,* November 1, 1965.

Chapter Eleven *A Backward Glance*

1. Cisler said there had been several offers, and he listed the arguments against selling in *Letter to the Board,* February 15, 1966.
2. As early as the board meeting on April 15, 1957, Cisler had discussed the costs and problems of the hydro plants. On September 21, 1959, the board authorized dismantling of the plants, and the offering of the properties for sale.
3. *Letter to the Board,* October 12, 1963.
4. *Synchroscope,* November, 1959. See also, The Detroit Edison Company, *Annual Report,* 1962.
5. The Detroit Edison Company, *Annual Report,* 1961. *Synchroscope* reported the completion of the closedown, July, 1962.
6. Service went to 725 residences, to 1952 business customers, and

with 16.5 miles of mains it was the largest privately owned water system in the state. *Synchroscope,* April, 1962.

7. Sales Department *Annual Report,* 1960.
8. For so small a transaction, negotiations were complicated. The Company must first deal with the board of the Thumb Cooperative, and then with its consultants. The matter then went to Washington. The Company proposal, in substantially the form in which it was finally adopted, was reported to the board November, 1961. Washington approval was reported in the *Letter to the Board* October 12, 1965. Edison revenue under the contract was estimated at $300,000 per year.
9. *Letter to the Board,* June 7, 1964.
10. Cisler to Jerome P. Cavanaugh, March 6, 1968.
11. A further addition to this complex followed the dismantling of the old Orchard Street d.c. station adjacent to G. O. building. A new building housing the communications center was erected on the site.
12. *Synchroscope,* June, 1950.
13. The committee consisted of Fred Compton, George Porter and Charles Landrigan.
14. *Synchroscope,* February, 1959. The work of rebuilding and reequiping had begun July, 1956.
15. Chester F. Ogden to Cisler January 24, 1961.
16. *Minutes of the Board,* January 18, 1954.

PARTS THREE AND FOUR

There is a proverbial reference to the amount of paper it takes to run an army; a great corporation must share that distinction. Given the size and the diversity inside the Company, it is not surprising that the sheer bulk tended to be overwhelming.

There was a tendency, perhaps natural in an enterprise so dependent on technology, to discard material which was no longer relevant to current operations. Retirements and changed assignments often invite a disposition of records.

The problems discussed in Part Three are those of the contemporary generation within the Company, and the men who were party to the decision-making process are often still within the Company. With a generosity which I much appreciated, these men made available to me on request the relevant materials from their respective offices. Let me testify that a great many records, minutes, and extensive correspondence survive.

One highly significant collection must receive special mention. Literally tons of paper have flowed into and out of the office of Walker Cisler in the almost twenty years in which he has served as President and/or Chairman of the Board. Materials classified as current occupy ten four drawer files in his office; noncurrent and inactive files from earlier years have usually been discarded.

This material is well organized, and the finding list furnished to me by Miss Evelyn Lehman, Administrative Assistant, allowed a thorough but discriminating search of these files. They are especially full in the questions involving intercompany matters, relations with professional or political authorities, and in the public and civic activities of Cisler personally or in the name of the Company. The study of the recent past, or the analysis of current plans, would have been quite impossible without this access.

Chapter Twelve *The Imperatives, and the Decision Making Process*

1. Reprinted in the *Detroit Free Press* and distributed in this form with the *Letter to the Board*, April 14, 1964.
2. The Detroit Edison Company, *Annual Report* for 1966.
3. Election of officers usually occurred at the time of the annual stockholders' meeting, but there was no rule to that effect, and exceptions were common.
4. The Detroit Edison Company, *Annual Report* 1964.
5. The resignation was apparently unexpected, and the date unusual. *Synchroscope,* June 1967.
6. Leahy in an address to the Business Club referred to the "inventory of management resources" and to the established necessity for a program of training for the replacements for the "startling number of men who must soon be replaced." *Synchroscope,* April 1963. In 1963 and 1964, a special retirement allowance was granted, often to the older men; Leahy made it quite clear that the purpose was to allow promotion of selected younger men in the departments.
7. The annual meeting of the shareholders was usually followed by a meeting of the directors. The minutes of the meeting of May 7, 1956 record the change in membership of the board to not less than seven or more than fifteen; the election of two new members increased the number from eleven to thirteen.

8. The problem was not unique to Detroit Edison or even to regulated utilities. Cisler included with the *Letter to the Board,* August 17, 1962, a copy of a speech made by Thomas R. Reid, civic and governmental affairs manager for Ford Motor Company. Reid had addressed the Edison Electric Institute, meeting in Atlantic City, June 6, 1962. The essential argument was that in this complex society, big government was both a fact and a necessity. He urged the acceptance of the mood of cooperation rather than continued hostility. The tasks which government was accepting, he insisted, were those which for some reason business had failed to accomplish.
9. This growth in income was translated into substantial increase in profits, and attracted the attention of the business world. *Forbes Magazine,* May 1, 1966 had a long congratulatory article on the Detroit Edison Company and its management. In part, the increase came from sales to the automotive companies, but Eldred Scott was careful to point out that the steel and chemical industries had also been factors.

CHAPTER THIRTEEN *Integration and the Altered Pattern*

1. *Letter to the Board,* January 11, 1961.
2. *Letter to the Board,* November 10, 1964.
3. Cisler as current president of the EEI was extensively quoted in the press release December 13, 1964. He emphasized the extensive development in interdependencies, which the *Survey* had tended to ignore. He especially urged the continued developmental method, as opposed to the "blue print" of the *Survey.*
4. The postwar situation is shown in the report to the Federal Power Commission, April 4, 1947.
5. The joint statement of the two companies as released to the press January 4, 1963, was signed by Walker Cisler on behalf of Detroit Edison, and by A. H. Aymond, Chairman of the Board, and James A. Campbell President, for Consumers Power. It emphasized the fact that there was no merger, but an agreement between the two companies, each of which would rank among the nation's largest ten.
6. Committee files contain the final draft, as well as some preliminary drafts, together with notations and criticisms.
7. *Letter to the Board,* January 19, 1963. To the employes, the emphasis was on the potential for development in the new scale. The general importance was noted in the *Annual Report* for 1962.

8. Both the *Wall Street Journal* (May 25, 1961) and the *Nations Business* (September 1961) were quite explicit in insisting that the leadership of the Department of the Interior, the Federal Power Commission, the REA and the TVA were all actively pushing for a system which would increase federal electric power production, and would establish federal control of transmission. In January, 1959, Senator Murray of Montana (Democrat) had introduced a resolution calling for investigation of investor owned power systems and their interconnections. The measure, S.R. 71, did not pass, but the incoming administration was expected to be more favorable to the purpose.
9. A bit of fencing occurred in 1959, when the Company received a report from the commission covering a study of possible interconnections southward across the Ohio line. The study, which the commission asserted had been made at the request of the Defense Department, proposed interconnection as a means of strengthening Michigan's reserves, in the national interest. The Company response attempted to avoid the charge of noncooperation, but was minimal and negative. The interconnection would have established federal jurisdiction, and this fact no one forgot. *Letter to the Board,* October 14, 1959. *Minutes* of the board of directors, October 1959.
10. *Letter to the Board,* July 2, 1963.
11. The proposal would repeal section 202F of the Federal Power Act as amended, which exempted international interconnections from FPC jurisdiction. The commission also sought authority to prohibit interstate interconnections, or, on its own authority, to order such interconnections.
12. A copy of Swidler's letter to Magnuson was sent by the Senator to Cisler; the board received copies.
13. Copies of the letters and of the Company response went to the board; Cisler told the board that he intended to consult Consumers Power and perhaps others. *Letter to the Board,* February 11, April 25, 1964.
14. The refusal, over the signature of W. R. Strike, chairman, was addressed to Wallace B. Brand, staff attorney for the FPC. Cisler received a copy from the sender. *Letter to the Board,* October 5, 1965.
15. Cisler explained the actions taken, and the reasons for Company actions, in his speech to the local shareholders. *Synchroscope,* December, 1965.
16. Quotations are from the law. The argument is in the order of the

commission, February 4, 1965. For Company reaction, see *Letter to the Board,* February 9, 1965.

17. The argument attempted to prove interstate interest from the fact of New York incorporation. At the request of the Company, the Michigan law was amended to allow dual incorporation. Michigan Public Act 1965, #161, Section 187a, approved July 12, 1965. By the time the Company was reincorporated, the controversy was passed, though it did clear the way for approval of some security offerings.
18. The statement of issues and intentions is from a memorandum from Paul McQuillen to Walker Cisler, June 4, 1966. Lee G. White of Nebraska, replaced Swidler, who did not wish reappointment. *Washington Post,* February 13, 1966.
19. Don Kigar, president of the Company, in his annual talk to the Power Club, sketched the construction program. *Synchroscope,* February, 1966. The first 345,000 volt interconnection with Consumers Power was closed December 10, 1968. *Letter to the Board,* January 14, 1969.
20. The interconnection between Consumers Power and the interstate companies was closed June 3, 1969. The interconnection southward from Detroit Edison toward Toledo was delayed as a result of a local dispute over zoning regulations. Construction was completed late in 1970.
21. *Letter to the Board*, August 15, 1967. It became operative May 30, 1969. *Letter to the Board,* June 10, 1969.
22. The reaction to the blackout was almost as confusing as the experience. Individual customers demanded assurance that this could not happen to them. The U.S. Congress was moved to investigate. Senator Cotton was sure that the interconnections themselves were to blame, and charged the industry with the sacrifice of security to cheapness. By contrast, the Electric Power Reliability Act proposed in 1967, would give to the Federal Power Commission broad powers over the construction and operations of generating facilities, and final determination over the ehv transmission systems of the nation—public and private. Cisler contrasted the concerns expressed with the 99.9 percent reliability which the industry maintained.
23. The Company shared in several local or regional associations; (1) the Michigan pool and center, (2) MIIO, (3) ECAR and (4) NERC. Each organization printed material which varied from publicity throw aways to technical articles; member organizations received minutes, committee reports, and special studies. The material is extensive, and it seemed unwise to attempt in so brief

an account to document the actions of the regional and national associations.

24. The contingency studies began in the summer of 1968, well before the MIIO connection was closed. The MIIO studies were submitted to the executive committee of ECAR, and on committee approval were furnished to the Federal Power Commission. *Letter to the Board,* March 12, November 12, 1968.
25. Before the blackout, the order of the commission (July 7, 1965) provided for advisory committees of its selection. The order of January 10, 1966 modified the committee structure to achieve the regional form.

CHAPTER FOURTEEN *The Production of Power*

1. The forecast by the Load Study Committee, for the years 1963–67 (May 11, 1963) was highly accurate for 1963 and 1964, but proved to be sharply below the realized demand which developed thereafter, both in peak demand and in annual production.
2. The major installations in the Michigan pool were listed in the planning stages. The table is compiled from the announcements in the *44th Electric Power Survey,* 1968.
3. The Capital Expenditures Committee, in its report for the year 1963, had anticipated no important expenditure for power production before 1967. *Letter to the Board,* January 15, 1963. The revised plans called for the expenditure of $400 million, over five years, an average of one quarter of a million dollars per day. The decision to build as soon as possible was given. *Letter to the Board,* July 15, 1964. In the *Letter to the Board,* September 15, 1964, Cisler gave the figures to show that the sharp upward trend called for a revision of the total planning by the industry.
4. *Letter to the Board,* July 15, 1964.
5. *Letter to the Board,* July 13, 1965.
6. *Synchroscope,* February, 1965. The details of the plans, including the discussion of the location, had been given in *Letter to the Board,* January 12, 1965.
7. *Letter to the Board,* September 11, 1967.
8. *Letter to the Board,* March 15, 1966.
9. *Synchroscope,* February, 1961.
10. *Synchroscope,* June, 1966. *Letter to the Board,* December 12, 1967.
11. Memorandum and Report, Wagner to Cisler, November 5, 1961. Also *Letter to the Board,* November 15, 1961.

12. Interest in these projects was high inside the Company. The public press was not always cautious and an occasional shareholder became alarmed over the threat of progress. Landrigan reassured one such individual with the assertion that "The Company. . . . will be ready to take advantage of every improvement in the art."
13. The U.S. Congress passed and President Johnson signed an Omnibus Bill, one provision of which declared that the old channel of the river was no longer a navigable stream. The action cleared the way for the utilization of the site.
14. *Letter to the Board,* August 15, 1966.
15. The board was informed of the action by Consumers Power, in approving the joint venture. The decision of the other companies mentioned, if and when their commitment became firm, could modify structure and scope. *Letter to the Board,* August 15, 1967.
16. Because of the unique elements in the project the public information departments jointly issued an elaborate pictorial description, giving specifications. An article in the *New York Times* (November 26, 1967) reported that seven more pumped storage projects were in construction, and forty-eight others were under consideration.
17. The press release of July 26, 1968 included both this Fermi II plant and unit # 3 at Monroe.
18. *Letter to the Board,* August 16, 1968; November 12, 1968. The general discussion of nuclear fuel and Fermi I is reserved for Chapter Sixteen.

CHAPTER FIFTEEN *Fuels and the Attendant Problems*

1. There was a constant threat of labor disturbance on the lakes. One firm, which delivered almost half the fuel, was not unionized. Two others were organized, and the effort to complete unionization could produce trouble for the St. Clair plant. When the usage was small, as had been frequently the case at the older plants, the costs of handling stored coal at these locations were high, and the shipping savings could be offset. *Letter to the Board,* September 14, 1960.
2. Robert W. Lundgren to the Power Club, March 13, 1969. Lundgren, formerly in charge of fuel procurement, gave the story in his address to this employe organization, and graciously made the manuscript available to me. His responsibility included negotiation with both the coal suppliers and the coal railroads,

and his report covered much of the decade which followed 1958.

3. *Letter to the Board,* November 16, 1960.
4. Substantial agreement had been arrived at in the meetings in Washington, June 16 and June 27, in which seven railroad presidents were involved, and the examiner for the Interstate Commerce Commission also. The final meeting on detail was scheduled for Detroit for July 17. *Letter to the Board,* July 12, 1961.
5. Printed by The Edison Illuminating Companies.
6. The negotiations, often on the personal and nondocumentary level, were described in the *Letter to the Board,* September 11, October 5, and November 13, 1962.
7. The negotiations with the Baltimore and Ohio Railroad involved shipments from the mines to Toledo coal docks. *Letter to the Board,* October 13, November 10, 1964.
8. *Synchroscope,* July, 1966.
9. These tariffs went into effect January 1, 1968. At that time, Cisler estimated that the savings from the three tariffs and the pipeline tariff would amount to $14 million a year.
10. Walker Cisler went to the board with his formal recommendation on December 23, 1966, for approval at the next meeting. He enclosed the copy of the memorandum dated December 21, and signed by Eldred Scott, George A. Porter, Harvey Wagner, and Marshall Pease. Scott had drafted the recommendation for the use of coal. "The above judgment is based, of course, on many assumptions and projections. It is no better than they are. Yet. . . . it seems prudent that we choose the alternative with which we have had more experience, and especially since it involves the lowest capital outlay."
11. The joint announcement took the form of a press release, December 18, 1968.
12. *Letter to the Board,* May 13, August 13, 1968.
13. *Synchroscope's* "Billion Dollar Story" March, 1968, also issued as a separate publication by the public information department.
14. *Synchroscope,* July, 1969.
15. *Letter to the Board,* November 12, 1963.
16. The board was informed that the requirements set by city and county regulations could be met by the expenditure of $4.5 million. *Letter to the Board,* March 15, 1966. By the end of 1966, including the work to be done on the industrial plants taken over, and on those plants outside Wayne County, the estimate was up to $10 million.
17. Lundgren to Kigar October 10, 1967.
18. When Senator Muskie's bill on control of air pollution was before

congress, Harvey Wagner was invited to testify, especially on the Company experience with the dolomite process. His testimony was that this method, combined with scrubbers, might be effective, but that the cost was unknown and the process untested. The experience of the Company in a test use of the dolomite method at St. Clair proved it to be ineffective in the form used, and troubled with side effects. *Synchroscope,* December, 1967.

19. In an elaborate and thoughtful address to the state wide Conference on Air Pollution, Michigan Chamber of Commerce, October 1, 1969, Wayne L. Wingert detailed the experiences of the Company in the efforts at control of sulphur oxides, and concluded that "the tall stack is the only feasible answer . . . for use today."
20. *Letter to the Board,* June 10, 1969. On January 19, 1969, the international joint commission held a meeting in Windsor to examine the effectiveness of the programs of the United States and Canada in abatement of river pollution. At that time, thermal pollution was not thought to be a problem. A subsequent meeting was called to discuss temperature criteria. *Letter to the Board,* February 10, 1969.
21. *Synchroscope,* June, 1967.
22. In an address Wingert listed the plants one by one, discussed the nature of the problem each presented, and, using another summary of costs, listed expenditures to date to a total of $62 million. The manuscript of the address to a Conference on Air Pollution Control, Oakland University, November 5, 1969 was made available to me by the speaker.
23. The members of the Environmental Advisory Committee were identified by Harvey Wagner. The Company was especially concerned with the accuracy of the monitoring devices used and with the acceptance of the findings by outside agencies.

CHAPTER SIXTEEN *Fermi and the Future*

The technical problems of the production of electric energy by atomic fission are obviously beyond the scope of this volume. The expert may well be disappointed in the treatment of matters of his concern. In this chapter, I have attempted to put the story of Detroit Edison's adventure in the field of atomic power into its proper prospective, as a part of a total history, and make understandable the purposes of the men who made the decisions. The discussion of nuclear technology is carried only so far as is necessary to achieve these aims.

Special reference should be made to one valuable secondary work,

Grace M. Pacak, *History of Atomic Power Development Associates, Inc., 1955–1965*, (typescript, n.d.) which was made available to me by the author. This is a well researched and detailed account of the corporation, with extensive material on background before 1955, and full information on the changing corporate memberships, on technical and other personnel, on research and publications and on finance. Miss Pacak was for some years an employe of the corporation, and had full access to its records. I am happy to express my deep appreciation to her for permission to use the manuscript. To my knowledge, no comparable history exists for PRDC, though I have heard of some work in progress. Mention should be made of the fact that the Michigan corporation, the Power Reactor Development Corporation, issued an annual report from 1956 onward. For organizational and financial facts, this report is continuingly useful. In some years, the discussion of events and achievements is quite extensive; even in years of misfortune, the report to the member corporations is straightforward and candid.

Dr. Richard C. Hewlett is a professional historian with the Atomic Energy Commission. On April 27, 1967, he delivered a paper at a joint meeting of historical associations in Chicago on the subject "Industry and the Atomic Energy Commission, 1947–1954." I wish with thanks to acknowledge his courtesy in allowing me to use the manuscript of the address. It is a careful historical narrative, based on the records of the commission, and directed to the problem of rewriting the law. It was especially helpful to be able to examine the Company's own records of this period, in the light of this study of the commission's problems and purposes. For the earlier part of the story, the relevant records of the Company are in Records Center.

1. The formal report of the committee to the Association of Edison Illuminating Companies was dated October 4, 1946, and bears the title "Atomic Energy for Power Production." At about the same time, the Committee circulated a press release from the War Department, dated November 22, 1946, with the comment that it confirmed the committee judgment of the expense and difficulty of any significant development. With the minutes and the report, there is extensive correspondence on Atomic Energy in the Records Center.
2. Richard G. Hewlett, *op. cit.*
3. McCullough wrote from the Clinton Laboratories, Knoxville, to issue the invitation. C. Rogers McCullough to Parker, November 21, 1946. The correspondence, in the effort to get clearance, extends to July 28, 1947, when W. B. Daume on behalf of Monsanto informed Parker of the actions taken; his contract "Consultant

subcontract #74 under prime Contract W35–058–eng–71" would accordingly not be renewed. This was not the end, however, for as late as June 22, 1958, Parker was still filing statements.

4. The report itself covered twenty pages, and was accompanied by a covering letter from Parker dated December 22, 1948. Lilienthal's response was a very thoughtful five page single spaced discussion of the issues raised. Press coverage was extensive, and included a press conference in which Lilienthal and Parker gave a joint interview, December 29, 1948.
5. The minutes of the final session show that the chief discussion was whether, given the circumstances, it was worthwhile to meet again.
6. The letter of intent informally divided the responsibilities of the two companies. Mark B. Putnam, vice-president for Dow Chemical Company, to James W. Parker, president, The Detroit Edison Company; the letter was countersigned by Parker in acceptance.
7. The public announcement of the creation of the department, and of its leadership, was in *Synchroscope,* July, 1952.
8. The extensive correspondence leaves no doubt of the interest of the Atomic Energy Commission in this private venture, or of the willingness of the staff to share in planning it. There is also extensive correspondence with other corporation officers on possible participation, 1952 ff.
9. As of June 11, 1953, the list of persons cleared for handling confidential material included more than three hundred, from 26 different business organizations.
10. For Cisler, this had already taken on the character of a crusade. In his address at a June, 1954 convention in Atlantic City, he had stressed the rising trends in demand for energy. Already the United States accounted for 40 percent of all the power used in the world; projection of trends made the development of atomic power necessary in the public interest. *Synchroscope,* July, 1954.
11. This controversy and its meaning are discussed by Richard G. Hewlett, *op. cit.*
12. The headlines about cheap energy and the charges hurled roused some fears about the future of privately owned utilities. Eldred H. Scott, of Detroit Edison, presented a discussion of the "probable impact of Atomic Energy on Electric Public Utilities Securities" and concluded that there would be none. *Journal of Finance,* May, 1954.
13. The original highly informal letter of agreement between Dow and Edison was superseded by the more formal agreement, to which the Atomic Energy Commission was a party, dated March

23, 1951. These original agreements were modified at least twice before the August 31, 1954 separation.

14. The action on behalf of The Detroit Edison Company is reported in the minutes of the board, December 15, 1954. There was an intermediate organization of associates after the withdrawal of Dow and before the incorporation.
15. The Detroit Edison Company, *Annual Report,* 1955, gave the full detail of the corporate structure and of the plans as they had so far been developed for "Lagoona Beach." For legal purposes, the separation was complete; in the *Reports* of the Company for the years 1956 and following, the extensive account of the atomic power program was made separately and independently. With Cisler serving as president of both corporations, however, and with personnel on temporary loan, the separation was not in fact complete.
16. Cisler was the presiding officer at the dedication. Dan Karn, of Consumers Power, and vice president of PRDC, made the name presentation. The story, with pictures, is in *Synchroscope,* September, 1956.
17. The Cisler statement of May 24, was printed in *Synchrocope,* June, 1956.
18. The report and the financial decisions were made at the meeting of the board on October 13. The approval of the commission was noted in the board minutes for November 19, 1956.
19. *Synchroscope,* January, 1956, noted the three sets of hearings at the same time; in this and other issues, the magazine gave coverage to the hearings.
20. The quotation from Judge Burger was given prominence in the *Wall Street Journal,* November 15, 1960. Cisler also quoted some of his dissenting opinion in *Letter to the Board,* June 15, 1960.
21. *The Detroit News,* June 14, 1960.
22. *The Wall Street Journal,* August 15, 1960, discussed the Washington politics involved, along with the legal issues in the appeal.
23. The rather involved arrangements for the joint pleading of the Solicitor General on behalf of the United States, of the Atomic Energy Commission and of PRDC were explained in *Letter to the Board,* November 16, 1960.
24. The first such statement was July 1, 1953, when Cisler reported for the Dow Chemical-Detroit Edison Associates. On May 10, 1954, he reported for the new organization, Atomic Power Development Associates. On February 16, 1956, the report was on behalf of the Power Reactor Development Company, a style which continued. The series is complete in Records Center.

25. The form of the committee's report is itself interesting. The adverse report of an earlier date was reintroduced and, point by point, the committee pronounced itself as now satisfied.
26. For a number of years, Arthur Griswold had been staff man inside Detroit Edison charged with the responsibility for direct relationship with the Fermi project. His election as vice president and general manager of Fermi (on leave from Edison) indicated the expectation of more usual business relationships and problems at Fermi.
27. In June, 1965, the public information department of Detroit Edison issued a new booklet, "Atomic Power for Peace and Prosperity," replacing older material for use at the Atomic Information Center at the site.
28. At the Third International Conference on the Peaceful Uses of Nuclear Energy, held at Geneva, the Detroit group had an exhibit on the breeder reactor. Cisler and Griswold were there, with some others from the Company; in attendance were 3,500 persons from 73 countries. *Synchroscope,* December, 1964; *Letter to the Board,* September 15, 1964.
29. Both the *Detroit News* and the *Detroit Free Press* carried extensive stories August 8, 1966. Cisler was interviewed, and the papers were generous in their congratulations. Copies went to the board on August 15.
30. The importance of Fermi, and the significant place it held worldwide in the development of nuclear power, are both reflected in the expressions of sympathy and in the offers of assistance in the emergency. The Chairman of the United Kingdom Atomic Energy Authority, Mr. C. M. Hill, wrote; "May I say how much we have appreciated the close cooperation which has been established between us over the past twelve years. I hope that we can be of real help to you in your present difficulty." Hill to Cisler, May 8, 1968.
31. Public Law 88–489 was signed by the president on August 26, 1964, as an amendment to the Atomic Energy Act of 1954. The provisions became mandatory at specified periods over several years. Cisler discussed the law's importance in *Letter to the Board*, September 16, 1964, and he was quoted publically in *Synchroscope,* October, 1966.
32. This was not a new problem, nor had it been previously ignored. Landrigan, in his statement to the Joint Committee on Atomic Energy, March 2, 1961, reported that the Company and four other utilities had joined the Davison Chemical Company in a feasibility study, and then in a much more complicated economic study.

From that date, there were continuous cooperative efforts in which Harvey Wagner frequently was Company representative.

33. The fact should be emphasized that the critical decisions were the result of changes in the national situation, and that they date from well before the accident of 1966.
34. Parallel with PRDC in time, other groups using other techniques built plants using nuclear energy. These groups frequently had interlocking memberships and, with rare exception, the member companies were or had been members of APDA or PRDC or both. Detroit Edison shared in one or more of these other projects, perhaps most importantly in the Peach Bottom plant of Philadelphia Electric.
35. There is no problem in the access to the financial figures. As a Michigan corporation, Power Reactor Development Company issued a printed annual statement, the first in 1956. The listing of memberships and contributions is complete. The reports to the Atomic Energy Commission and to the Joint Committee also carry required financial information. Cisler, of course, discussed the total situation and the involvement of Edison in it in *Letters to the Board.*
36. When Fermi received permission from the commission to resume testing, costs were incurred beyond funds then available. It was still the belief that Fermi, if operated as a research and service facility, could be made self supporting. Detroit Edison made a pledge of $3.5 million, payable over two years, and the Central Research Institute of the Electric Power Institute of Japan pledged $1 million in the effort to achieve this objective.
37. Not included in this figure of 1968 is the amount pledged for 1969–70, which has been mentioned.
38. C. Hoyt Whipple, professor of Radiological Health in the University of Michigan, speaking at a conference (March 11–14, 1970) in Ann Arbor on the subject, "The Threat of Nuclear Power," worded his argument sharply. "Nuclear energy is the cleanest, safest and largest energy resource we have and it has come along just in time.... The threat of nuclear power is that public opinion will delay the acceptance of this form of energy until we have wasted our reserves of fossil fuels and fouled our world. . . ."

CHAPTER SEVENTEEN *The Corporate Citizen*

1. The policy had long standing, but was publicly restated in detail in response to an employe question. *Synchroscope,* March, 1959.

2. In 1963, Cisler formulated this doctrine of responsibility into a statement which appeared in paid notices in the press, and was reprinted in the *Synchroscope,* November, 1963. The statement was intended "to express the responsible character of organization." The final principle announced the intention "to participate in and support economic social and cultural endeavors that advance the best interests of our customers and our communities."
3. The Detroit Edison Company, *Minutes of the Board,* January 20, 1958. Robert E. Schwab reported that the budgeted amount for the coming year was $157,000 for gifts. Schwab to Cisler January 11, 1960.
4. The contributions to all causes including membership, was $837,000, which was .86 percent of income before taxes.
5. The Capital Funds Committee was at first the expression of the civic concern of Frederick C. Matthaei and Selden B. Daume, who conceived of it as an instrument directing contributions from business enterprises to the best public and civic purposes. It was incorporated May 20, 1951. The minutes report on the organization, and on its subsequent activities over a decade and a half.
6. The Detroit Conference on Civic Development was an incorporated body; the minutes of both the meetings of the board and of the membership were available to me.
7. Detroit Edison Company, *Annual Report,* 1958.
8. The mayor of the city of Detroit to whom reference was made was Louis Miriani.
9. In December, 1964, the Detroit Conference on Civic Development issued a small pamphlet to celebrate its ten year history. At that date, the directors of the organization were Walker Cisler, Selden B. Daume, Ruth E. Stevens, Carl Johnson, John L. Copeland, Very Rev. L. V. Britt, S.J., M. A. Cudlip and Joseph Ross.
10. When these activities became statewide, it is sometimes difficult to identify Company participation. Sometimes the association is clear. The census of 1960 revealed the plight of the Upper Peninsula, and when he was a visitor there, an appeal was made to Cisler for assistance. Prentiss Brown was a resident of the area, and gave support to this activity, so far from Edison's lines. Company personnel gave assistance to the organization of "Operation Action-U.P.," which proposed to stimulate and direct local activity in a program of self-help. The Detroit Edison Company, *Annual Report,* 1964.
11. One simple index is the number of requests by demolition contractors for the removal of Edison equipment from houses. In

1964, the requests numbered 5,500; in 1965, there were 4,545; and in 1966, 1,100 such requests. These included both slum clearance and expressway construction.

12. Cisler was frankly delighted to have secured Doxiadis for the project; "He is probably the foremost urban planner of this generation." *Letter to the Board,* August 19, 1964.
13. The completed publication was as follows:
    Emergence and Growth of an Urban Region. The Developing Urban Detroit Area.
    Volume One; *Analysis.* Detroit, 1966.
    Volume Two; *Future Alternatives.* Detroit, 1967.
    Volume Three; *A Concept for Future Development.* Detroit, 1970.

    Not a part of the original plan for the series, but a special response to the 1967 summer crisis, was the much less elaborate, typescript, paper bound publication *A Course of Action* (Detroit, September, 1968). This publication was in part superseded by Volume Three.
14. Cisler's first report to the board was in the *Letter* dated August 7, 1967. One employe had been injured, though not seriously and not by a rioter. The area over which some damage had been done to Company property was some 40 square miles. Earlier in the year, the planned procedures for civil defense had been revised, and were put into use here. The activity of the department and of the Company in the crisis and in the subsequent recovery period is included in the report of the department on the subject, dated August 15, 1967.
15. *Letter to the Board,* October 10, 1967.
16. Employment of the disqualified could be dangerous in the Company's service area and a hazard to the worker. The Company accepted a contract with the Department of Labor, by which 120 such marginal workers were accepted for a special training program over two years. The Company received $3,000, approximately, in partial reimbursement for the high costs of training and counseling. *Letter to the Board,* March 10, 1969.
17. The phrase "The Detroit Edison and the Urban Challenge" appeared twice. It was used as the title to an article in the *Synchroscope,* September, 1968, which in an elaborate picture story featured Company efforts. It was used also in a public information pamphlet, December, 1969, which described the "participation in programs concerned with the developing human and economic resources of the Metropolitan Detroit area." The activities listed included those already mentioned, the MDCDA and

the New Detroit Committee. Beyond these were eighteen programs or organizations briefly discussed and others only mentioned in which the Company had participated in major or minor degree. In the language of this report, "involvement may include financial support, senior executive contributions of time and leadership to committees or boards, contributions of staff time, office space, commercial and technical services, consultative services, employment and training of disadvantaged members of society, part time employment of high school students, and cooperation with numerous private and public agencies who refer applicants to us."

18. The first public announcement occurred at the office of Mayor Roman S. Gribbs, May 7, 1970. The Mayor expressed satisfaction that development on such a scale was under way, and reinforced the implied invitation to other corporations to join in the effort.

Chapter Eighteen *On the International Scene*

1. Cisler's recommendation of the energy survey led to the invitation to help. ". . . . I have been requested to assist in getting such a survey under way." Throughout the year, Cisler reported on development to the board. The quote is from the *Letter to the Board,* December 17, 1967.
2. "It is my feeling, that because of what we have been doing in the matter of energy and power with other countries that it has been helpful in keeping the U.S. in a more favorable position than we are in other international respects." *Letter to the Board,* September 17, 1965.
3. *Synchroscope* (December, 1962) carried the story of the one thousandth delegation from Japan. Harvey Bumgardner was for many years the personal representative of Cisler in welcoming these visitors from abroad.
4. The high pressure pipe, involving extensive hand labor, was not produced commercially in the United States, because of the small demand; the order was divided between Japan and Germany. The purchase of the 20,000 kw machine for Beacon Street Heating Plant was approved by the board at a price substantially below the best American bid. *Minutes of the Board,* September 16, 1957.
5. In 1968, bids had been sought for the tower steel for 22 miles of 120,000 kw lines. In this case, the invitation to bid went to four

domestic and three foreign suppliers. *Letter to the Board,* April 4, 1968.

6. The reverse of this visitation was also important. Both the Department of State and the Department of the Interior on occasion requested assistance when Russian visitors were involved. Cisler was well known in the Soviet Union, especially among his fellow scientists, and their requests to Washington were both natural and frequent. Referring specifically to the visit of Vice-Chairman Novikov, of the Council of Ministers, who had been his host in Russia in 1959, Cisler said "I believe these reciprocal visits are constructive in bettering relationships between these two countries." *Letter to the Board,* June, 1964.
7. The Constitution of the World Energy Conference as amended was printed in London, January, 1969.
8. "This is the common goal for all of us. This is the first age that has ever considered the possibility of eliminating poverty, hunger and cultural deprivation in every region of the world. It is worthy of the best in us." The address, delivered at Basel, October 6, 1969, was entitled "New Horizons in Energy Resources for Mankind," and was given before *Nuclex 69.* Cisler was introduced to the audience as Chairman, International Executive Council, World Energy Conference; in the address, he identified himself with that organization.
9. "This honor, to the distinguished American utility engineer and executive is one in which the entire electrical power industry in the United States can take pride. . . . Crowning highlight of Walker Cisler's six year term as head of this organization will come in 1974 when the 8th Plenary Session of the World Power (cit) Conference will be held in the United States. We can conceive of no more fitting recognition than this for Walker Cisler's contribution to international progress and good will in the field of electric service." *Electrical World,* September 11, 1967.

Index